Vitactions

The New Vitamins for Well-Being and Healthy Aging

Rafael Franco

Title: **Vitactions**

The New Vitamins for Well-Being and Healthy Aging

ISBN: 979-8-89966-360-4

Author: Rafael Franco

Cover image: Rafael Franco

Publisher: Generis Publishing
Online orders: www.generis-publishing.com
Contact email: info@generis-publishing.com

Generis
PUBLISHING

Vitactions for Well-Being and Healthy Aging

Author: Rafael Franco

PREFACE

Observing as a Vital Action

This book proposes a simple and transformative idea: just as the body needs vitamins to maintain balance, the mind needs *vitactions* to sustain its vitality, clarity, and capacity for connection. *Vitactions* are small but essential actions that nourish attention, wonder, presence, and intention. The word "Vitactins" can be considered as a synonym.

"*Vitactions*" and "*vitactins*" appear frequently throughout the text because there are currently no widely accepted alternatives. When appropriate, we also use terms such as essential practices or essential habits to refer to them.

The author proposes using "*vitactin*" to refer to an individual action, in parallel with the way the term "vitamin" refers to a specific nutrient.

Throughout these pages, you will find images that accompany the concepts, stories, and reflections. They are not there merely as visual support, but as an integral part of the book's experience. Observe each image slowly. Allow yourself to pause for a moment, look at it with curiosity, notice what it awakens or suggests to you. Attentive observation is a door we often forget to open.

In a world where we gloss over almost everything, truly looking is a *vitaction* in itself. This book invites you to rediscover it: to read with presence, to see calmly, and to engage with every detail, without haste and without expectations.

May this journey accompany you as a form of inner nourishment, as necessary for the mind as vitamins are for the body.

Index

Chapter 1. Introduction – A New Paradigm

1.1 The Mind-Body Interface

Our behaviors are at the core of human health and well-being. They form the foundation of how our minds and bodies interact to maintain homeostasis and adapt to challenges. Over the years, research has demonstrated that certain behaviors are essential for physiological and psychological health. They are not merely lifestyle choices; they are fundamental requirements for optimal functioning. Yet, the consequences of their absence have been largely overlooked in modern healthcare. This book explores a recently introduced concept, "*vitactions*", used to describe these indispensable behaviors; for the reasons outlined below, the term *vitactins* is also used as an alternative. Vitactions/vitactins are actions that, when deficient, can lead to a cascade of psychosomatic imbalances. Unlike other frameworks that focus on diet, exercise, or isolated lifestyle factors, the concept of vitactions encompasses a broader understanding of behaviors as integral to our evolutionary and biological heritage.

1.2 Our Heritage as Humans

In our evolutionary past, humans developed specific behavioral patterns to survive, thrive, and adapt to their environments. Physical activity, social interaction, problem-solving, and exposure to nature were once inextricably linked to daily life. Such actions not only shaped our cultures and societies but also became deeply embedded in our physiology, influencing brain health, immune function, and overall wellness.

Longer life expectancy combined with modern lifestyles results in prolonged sedentary periods, social isolation, and a disconnection from nature. These shifts have produced a mismatch between the conditions our bodies evolved to depend on and those we now encounter. The consequences are stark: a rise in chronic diseases like type 2 diabetes, cardiovascular disorders, anxiety, depression, and neurodegenerative conditions.

The concept of vitactions arises from the recognition that reintroducing these essential actions into daily life is crucial to restoring balance, preventing disease, and achieving healthy aging. It provides a structured way to identify and address behavioral deficiencies that contribute to the growing burden of modern health issues.

1.3. Drawing an Analogy: Vitactions and Vitamins

To further clarify the significance of these two new terms for essential behaviors, the book draws an analogy with vitamins. *Vitactin* echoes the sound and framing of *vitamin* while retaining the ideas of vitality and action, giving the term a phonetic appeal that underscores the proactive, dynamic nature of these behaviors. The book uses *vitactions* and *vitactins* to refer to the same concept, although *vitactins* may resonate more strongly, especially when the analogy to vitamins is emphasized.

Just as vitamins were discovered to be essential for preventing and treating diseases caused by nutritional deficiencies, vitactions represent behaviors that are vital for maintaining mental and physical health. Whereas vitamins are chemical compounds required for biological processes, vitactins are behavioral stimuli that activate critical physiological and psychological pathways.

The analogy helps highlight the parallels between two essential pillars of health: nutrients for the body and *nourishing patterns* for the brain. In fact, vitactions are a distinct and innovative concept, built on the understanding of human biology and the demands of modern life.

1.4 Why Vitactions Matter Today

The rapid pace of urbanization, technological advancement, and lifestyle changes has left humanity at a crossroads. The decline in essential actions once embedded in everyday life has contributed to a host of *lifestyle disorders*. These conditions are not merely the result of individual choices but are symptoms of a broader societal shift.

By framing them as essential for health, much like vitamins are for nutrition, the concept of vitactions offers a practical roadmap for addressing these challenges. It emphasizes prevention over treatment, encouraging individuals, communities, and healthcare systems to integrate specific vitactins into daily life.

1.5 Goals of this Book

This book aims to:

1. Introduce the concept of vitactions, first coined in 2024 by Watve and Keskar Sardeshmukh, and establish its relevance in modern healthcare and well-being.

2. Explore the physiological and psychological pathways influenced by these essential deeds.

3. Draw an analogy between the early discovery of vitamins through deficiency syndromes and the emergence of vitactions through modern deficiency states driven by the loss of essential habits.

4. Provide actionable strategies for integrating vitactions into personal routines, community health initiatives, and medical practice.

Through this lens, *Vitactions for Well-Being and Healthy Aging* aims to inspire a paradigm shift, one that moves beyond simply managing symptoms to addressing the behavioral roots of many of today's most prevalent diseases. This is not merely a concept for academics or healthcare professionals, but a practical guide for anyone seeking to restore vitality and cultivate wellness in the modern world.

Chapter 2. Defining Vitactions and Vitactins

2.1 Describing Vitactions

In the vast landscape of health and a fulfilling life, behaviors often sit in the shadows of more tangible factors like diet, exercise, and medical interventions. Yet, these actions are far from trivial. They are essential to the intricate balance that sustains life, shaping both our mental and physical health. The term vitactions encapsulates this concept, describing something vital for the maintenance of physiological and psychological balance.

Vitactions represent the actions our ancestors performed routinely, movements, interactions, and decisions that became encoded into our genes and our physiology over millennia. Our physiology evolved under conditions where these behaviors were routinely required, so we are adapted to—and expect—them. Much like nutrients, they are essential for homeostasis and resilience. Their absence, referred to as vitactin deficiency (*avitactinosis*), can disturb this balance and contribute to chronic health issues. They are not optional habits or preferences; they are integral to human biology, deeply rooted in the evolutionary journey.

2.2 Key Characteristics of Vitactions

Vitactions are distinguished by their evolutionary, physiological, and essential nature. They are deeds that transcend culture and time, offering universal benefits to human health.

Over millennia, humans adapted to environments that demanded certain actions: walking long distances, social interaction for survival, complex problem-solving, and connection with nature. These practices became integral to health, supporting survival in the face of constant challenges. In the modern era, however, many of them have been sidelined by urban lifestyles and technological advances.

Vitactions are physiological triggers, not mere habits. Physical activity, for instance, promotes angiogenesis and enhances neuroplasticity, the brain's capacity to adapt structurally and functionally. This adaptability occurs even at anatomical and molecular levels, influencing synaptic remodeling, neurotransmitter balance, and the expression of genes involved in learning, memory, and emotional regulation. Social interaction releases oxytocin, reducing stress and inflammation. Vitactions act as catalysts to maintain cellular health, supporting mitochondrial efficiency, balanced glucose metabolism, and proper functioning of the immune system.

The hypothesis of behavioral deficiency syndrome posits that the absence of essential actions creates physiological and psychological voids. Much like nutrient deficiencies causing diseases, the lack of vitactions may lead to metabolic disorders and chronic conditions affecting the Central Nervous System, including depression, anxiety and cognition.

2.3 Vitactions in a Modern Context: A Behavioral Response to Today's Challenges

Modern lifestyles starkly contrast with the behaviors our bodies evolved to rely on. Ancestral environments demanded survival actions like physical exertion, problem-solving, and social bonding. Today, convenience, digitalization, and urbanization have replaced many of these needs. Physical activity is substituted with sedentary routines, social bonds with superficial digital communication, and cognitive challenges with passive consumption.

This mismatch is a primary contributor to modern health challenges. Chronic diseases, lifestyle disorders, and mental health issues are not solely the result of individual choices but symptoms of a broader behavioral deficit. Addressing these gaps through the lens of vitactions offers a way to reconnect with actions that sustain human health.

2.4 Preliminary Classification of Vitactions

Vitactions may be classified according to their areas of impact. A preliminary framework is presented in **Table 2.1**.

Category	Examples	Primary Benefits
Physical Activity	Walking, running, resistance training	Cardiovascular health, neuroplasticity, metabolic balance
Social Interactions	Group activities, emotional support	Mental well-being, stress reduction, immune system boosting
Cognitive Stimulation	Learning, puzzle/problem-solving	Enhanced cognition, delay in neurodegeneration
Nature Exposure	Forest walks, sunlight exposure	Mood enhancement, immune support, physical vitality
Risk-Taking/Adventure	Exploration, sports, challenges	Adaptability, resilience, dopamine sensitivity

Table 2.1. *An initial classification of vitactions, inspired by the works of Watve and Keskar Sardeshmukh and the author's own contributions (see references in the appendices).*

Physical activity is an essential vitaction. It includes walking, running, and other movements that mimic ancestral actions. Sedentary lifestyles characterized by prolonged sitting and minimal physical exertion are a "modern" deficiency. This lack of movement contributes to obesity, type 2 diabetes, cardiovascular diseases, musculoskeletal decline, and cognitive impairment.

Social interaction is another critical vitaction. Meaningful connections, emotional bonds, and collaborative activities are vital to human health. Isolation or superficial digital interactions represent a significant deficiency, leading to heightened risks of depression, anxiety, cognitive decline, and chronic inflammation.

Cognitive stimulation involves engaging in activities like problem-solving, learning, and creativity. Passive consumption of content or lack of intellectual engagement is a growing issue, resulting in cognitive stagnation, reduced neuroplasticity, i.e., the ability of our brain to adapt to a changing environment, and an earlier onset of dementia.

Exposure to nature provides unique benefits. Time spent outdoors, in forests, or near water bodies offers mental and physical health boosts. Urban lifestyles often limit this interaction, leading to mood disorders, vitamin D deficiencies, weakened immune responses, and general declines in vitality.

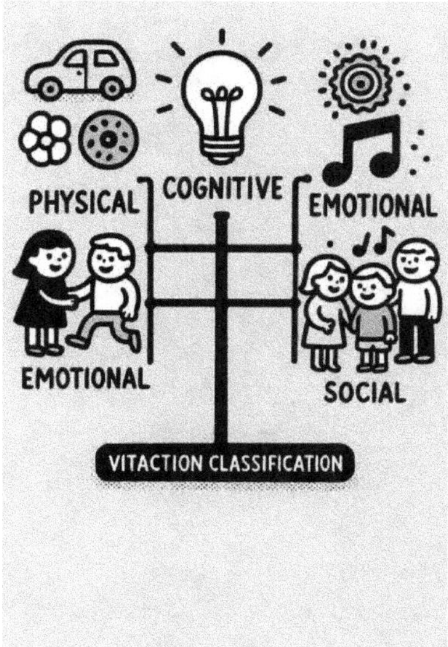

Risk-taking and adventure are also essential vitactions. They build resilience and adaptability through calculated risks or novel experiences. The absence of such challenges can lead to reduced adaptability, diminished dopamine sensitivity, and a weakened sense of purpose.

2.5 Conclusion

Vitactions/vitactins, offer a transformative way to understand health. These actions are neither luxuries nor optional; they are essential and deeply embedded in our evolutionary heritage. The hypothesis that deficiencies in these foundational practices contribute to modern health challenges opens new avenues for prevention and wellness. Moving forward, the focus will shift toward exploring the physiological mechanisms that link these actions to human health.

Chapter 3. Vitactions: Evolution, Mechanisms and Role in Well-Being

Human well-being is deeply tied to the activities we pursue, much like the physiological systems of the body depend on nutrients for proper functioning. Just as vitamins are essential to the health of organs and tissues, vitactions are indispensable for maintaining the conditions that support emotional stability, cognitive function, and social harmony.

As vitactions are not optional but evolutionary necessities, they serve as triggers for mechanisms that sustain balance and resilience in our interconnected minds and bodies. However, as modern lifestyles diverge from ancestral patterns, the absence of these essential activities has led to functional deficits, contributing to widespread emotional and cognitive disruption. This chapter explores the evolutionary origins of vitactions, the mechanisms they activate, and their indispensable role in supporting human vitality, resilience, and fulfillment.

3.1 The Evolutionary Nature of Vitactions

Vitactions are behaviors that emerged as fundamental survival strategies in the evolutionary history of humans. They are deeply ingrained in our biology because they addressed the basic needs of early hominids living in dynamic and often hostile environments. They not only ensured survival but also fostered the development of physiological and psychological systems that remain central to our lives today.

For millennia, physical activity was essential for survival. Hunting, gathering, escaping predators, and building shelters required endurance, strength, and agility. These activities shaped our cardiovascular, muscular, and skeletal systems. Regular movement also optimized energy utilization and improved cognitive alertness, as individuals who were physically active were better equipped to respond to environmental challenges.

Early humans lived in groups, and social interaction and cooperation were critical for survival. Emotional bonds, shared responsibilities, and collaborative problem-solving ensured protection, resource sharing, and the development of trust. Social interactions played a key role in shaping the social brain, fostering empathy, communication, and the ability to navigate complex relationships.

Adapting to new environments required cognitive stimulation, which consisted of problem-solving, toolmaking, and planning. These activities expanded cognitive capacities, laying the groundwork for innovation, learning, and strategic thinking. They also promoted neural adaptability, allowing humans to thrive in diverse and changing environments.

Immersion in natural environments was an intrinsic part of daily life. Sunlight regulated circadian rhythms and supported vitamin D synthesis, and diverse outdoor settings provided sensory engagement and stress reduction. Interacting with nature fostered physical health and emotional balance.

In ancestral environments these activities were essential for meeting the demands of life. However, in modern contexts, many of these vitactions have diminished or disappeared, leading to a disconnect between our evolutionary needs and the way we live today.

3.2 Vitactions as Triggers for Behavioral Well-Being

Vitactions serve as activators for biological processes that maintain emotional, cognitive, and social balance. They do not possess intrinsic mechanisms themselves but serve as triggers that activate key processes, ensuring that the systems governing emotional, cognitive, and social functioning operate at their best. Vitactions shape our brains: neural plasticity was triggered when early hominids practiced vitactions, and continues today whether we seek out these vital experiences or, on the contrary, fail to incorporate them into our lives.

Neuroendocrine regulation is a consequence of vitactin *consumption*. Vitactions influence the production of key hormones that regulate mood, stress, and motivation. Physical activity lowers cortisol, the body's primary stress hormone, and increases the production of endorphins and serotonin, which contribute to feelings of happiness and relaxation. Social interactions release oxytocin, fostering trust, reducing loneliness, and promoting emotional security. Likewise, cognitive stimulation lights up dopamine-driven circuits, boosting motivation and sharpening the drive for achievement.

Mitochondria are the energy powerhouses of our cells, and their efficiency is vital for maintaining physical vitality and mental clarity. Regular movement, as simple as walking or mild exercise, enhances mitochondrial function, ensuring optimal energy production and reducing oxidative stress. Recent research has linked mitochondrial dysfunction to neurodegenerative conditions like Alzheimer's and Parkinson's diseases. Vitactions that support mitochondrial integrity and functionality are therefore crucial for sustaining energy levels and cognitive resilience.

Neuroplasticity is closely linked to cognitive adaptability. The brain's ability to form new connections, respond to challenges, and recover from stress is fundamental to sustained mental performance and emotional resilience. Vitactions such as learning new skills, solving problems, or engaging in physical activity stimulate the formation of new neural connections. This process is supported by brain-derived neurotrophic factor (BDNF), a protein essential for neural health that enhances memory retention, learning, and adaptability.

Substantial evidence shows that vitactins regulate immune system functionality. Behavioral deficits, including prolonged isolation or inactivity, exacerbate systemic inflammation, which is linked to a range of mood disorders and cognitive impairments. Vitactions such as exposure to natural environments or participation in group activities may reduce inflammatory biomarkers, thereby enhancing mental clarity and emotional resilience and strengthening immune defenses.

3.3 Avitactinosis. Behavioral Deficits and their Consequences

When there is a shortage of vitactions, the mechanisms they trigger are left inactivated, leading to behavioral voids. These deficits can manifest in various ways:

- Emotional Imbalances: A lack of physical activity or social connection increases cortisol levels and disrupts mood regulation, leading to heightened stress, anxiety and/or depression.
- Cognitive Decline: Without cognitive challenges, the neuroplastic capability of our nervous system diminishes, accelerating the risk of memory loss and cognitive stagnation.
- Social Isolation: The absence of meaningful interactions leads to feelings of loneliness and reduces the production of oxytocin, exacerbating stress and reducing emotional resilience.
- Weakened Resilience: Prolonged behavioral deficits lower the body's ability to adapt to stress and recover from setbacks, creating cycles of emotional and cognitive vulnerability.

These deficits are not merely inconveniences; they represent significant barriers to maintaining a balanced and fulfilling life. Addressing these deficits calls for the intentional reintroduction of vitactions to restore the conditions essential for a rich and fulfilling life.

3.4 Synergistic Effects of Combined Vitactions

Each vitaction engages specific mechanisms; taken together, their effects can be even more powerful.

Outdoor physical activity combines exercise and exposure to nature. In fact, hiking in the mountains, running in the countryside or in an urban park combines the benefits of physical movement with the calming and immune-boosting effects of natural environments, amplifying stress reduction and emotional uplift.

Collaborative problem-solving has the advantage of social interaction. Group activities blend cognitive stimulation with social bonding. Team sports and creative workshops, enhance both intellectual involvement and emotional connection.

These synergistic interactions underscore the interconnectedness of vitactions and their potential to create cascading benefits across emotional, cognitive, and social dimensions.

3.5 Protective and Restorative Roles of Vitactions

There are different ways in which vitactions play both preventive and restorative roles in fostering emotional balance, mental adaptability, and social connection:

- In healthy individuals, regularly practicing vitactions helps prevent depression, anxiety, and cognitive decline by maintaining the mechanisms that uphold overall balance.
- For individuals experiencing behavioral deficits, reintroducing vitactions helps reverse disruptions, restore emotional stability, and rebuild cognitive and social resilience. Vitactions may also reverse metabolic diseases, at least in early stages. The most significant example is the reversion from hyperglycemia to normoglycemia in recently diagnosed type 2 diabetes patients who adopt regular physical activity and healthy nutrition habits.

Vitactions are therefore not only tools for maintaining balance but also powerful interventions for addressing existing imbalances.

3.6 Conclusion

Vitactions/vitactins are to the mind what vitamins are to the body: indispensable triggers that sustain balance, resilience, and vitality. A deeper understanding of their evolutionary roots and underlying mechanisms can lead to more effective responses to modern mental health challenges and promote ways of living that align more closely with our biological design.

The next chapter moves from theory to practice, exploring the key domains of vitactions and presenting strategies for integrating them into daily life.

Chapter 4. Categorizing Vitactions for Practical Use

Recognizing the value of vitactions is only the first step; what truly matters is having a practical framework to identify and integrate them into daily life. By organizing vitactions into distinct domains with concrete examples, individuals can align these vital activities with their personal goals and life contexts—nurturing emotional balance, cognitive resilience, and social fulfillment.

One compelling example can be found in activities related to music. Music itself is not a vitaction, but actively engaging with it—whether by listening, playing an instrument, singing, or identifying songs—serves as a powerful, multi-dimensional trigger for emotional, cognitive, and social enrichment. These activities illustrate the diverse ways vitactions can elevate human experience. This chapter explores their core domains and offers guidance on how to personalize them for lasting vitality, connection, and mental clarity.

4.1 Domains of Vitactions

Vitactions influence multiple dimensions of human behavior, which can be grouped into five primary domains. Each domain highlights a vital aspect, offering a roadmap for identifying and applying relevant vitactions.

The detailed categorization of vitactions, shown in **Table 4.1** with specific examples, helps to illustrate their practical application.

Domain	Category	Examples	Primary Benefits
Physical	Rhythmic movement	Dancing, rhythmic walking to music	Cardiovascular health, emotional uplift
	Structured exercise	Yoga, resistance training	Stress reduction, physical vitality
Cognitive	Active engagement	Learning to play instruments, guessing song names	Neuroplasticity, entertainment
	Creative stimulation	Composing music, writing, painting	Emotional expression, problem-solving
Social	Group collaboration	Singing in a choir, group drumming	Emotional connection, social bonding
	Cultural Sharing	Attending concerts, community music events	Shared experiences, collective joy
Environmental	Nature integration	Sunbaths, outdoor music listening, creating nature-inspired music	Stress relief, sensory harmony
Emotional	Mindfulness practices	Listening to calming music, focusing on melodies, focusing on lyrics	Stress relief, emotional stability
	Expressing creativity	Painting, playing or composing music	Emotional release, self-expression

Table 4.1. *Grouping Vitactions by Functional Category and Core Benefits*

4.1.1 Physical Domain

Vitactions in the physical domain involve movement and bodily activity, which are foundational for energy regulation, physical vitality, and emotional stability.

Examples:

- Walking or running: Simple, accessible activities that improve cardiovascular health and elevate mood through the release of endorphins.
- Dancing: Rhythmic movement, often accompanied by music, that combines physical activity with emotional expression.
- Structured Workouts: Activities such as Pilates, resistance training, or functional fitness exercises that promote strength, flexibility, and stress reduction.

4.1.2 Cognitive Domain

Cognitive vitactions immerse the brain in stimulating activities, promoting neuroplasticity and enhancing mental clarity. These practices foster adaptability and creativity, key components of cognitive resilience.

Examples:

- Learning to play an instrument: a practice that draws on rhythm, coordination, and memory to stimulate the brain.
- Analyzing musical pieces: an activity that combines critical listening, pattern recognition, and emotional interpretation.
- Playing strategy games: a practice that requires planning, adaptability, and problem-solving.

4.1.3 Social Domain

Social vitactions involve meaningful interactions with others, fostering trust, reducing stress, and enhancing emotional well-being. They are especially critical in combating loneliness and strengthening emotional bonds.

Examples:

- Group Singing: Joining a choir or participating in group musical performances combines vocal exercise with social connection.
- Shared Activities: Organizing a musical event or participating in a group music therapy session.
- Community Engagement: Volunteering or attending social gatherings where shared experiences build emotional bonds.

4.1.4 Environmental Domain

Environmental vitactions focus on engaging with natural surroundings, which reduce stress, stimulate the senses, and enhance emotional grounding.

Examples:

- Listening to Natural Sounds: Combining outdoor environments with attention to songs coming from birds, rain or wind.
- Paying attention to biosphere: Nature walks looking into the name, shape and movement of plants and animals.

4.1.5 Emotional Domain

Emotional vitactions regulate mood, build resilience, and provide a creative outlet for emotional expression. They are particularly valuable in managing stress and fostering a sense of calm.

Examples:

- Listening to Music Mindfully: Focusing on the subtleties of melodies, rhythms, and harmonies to evoke positive emotions or process complex feelings.
- Expressive Music Creation: Composing or playing music as a form of emotional release.
- Journaling Inspired by Music: Writing reflections or creative pieces as one listens to emotionally resonant music.

4.2 Behavioral Profiles and Personalization

Vitactions are most effective when tailored to individual preferences, goals, and circumstances. Behavioral profiles provide a personalized framework for selecting and prioritizing vitactions based on factors like age, lifestyle, and specific goals.

4.2.1 Key Personalization Factors

- Life Stage: Children are likely to benefit from participation in active and social vitactions, including dancing and group games, whereas older adults tend to benefit more from emotional and cognitive vitactions, including mindful listening or learning to play an instrument.

- Goals: For stress relief, emotional and environmental vitactions, like mindfulness practices or nature-inspired music, are particularly relevant. For cognitive enhancement, activities like solving puzzles or analyzing music may take precedence.
- Physical and Cognitive Abilities: Accessibility is crucial. For those with physical limitations, listening to music or singing may be more feasible than active movement or playing an instrument.

4.3 Strategies for Integration

Incorporating vitactions into daily routines requires flexibility, creativity, and intentionality. Effective strategies include:

- Habit Stacking: Seamlessly integrate vitactions into your daily life by pairing them with routines you already have. Practice mindfulness while listening to music after waking up, or turn your daily commute into a vitality boost by using it for intentional breathing, reflective listening, or focused attention.
- Incremental Engagement: Start with manageable actions, such as 5–10 minutes of mindful music or light movement, and gradually increase involvement as the habit becomes more natural and rewarding.
- Blending Vitactions: Amplify the impact by combining multiple domains, for instance, play an instrument outdoors or join a vocal group that performs in nature.
- Leveraging Technology: Using apps for interactive engagement.

4.4 Conclusion

Framing vitactions within functional domains offers a practical lens for understanding how they can be integrated into daily life. For example, engaging with music, through listening, performing, or singing, demonstrates how a single vitaction can activate multiple domains, enriching emotional, cognitive, and social dimensions of human experience.

By tailoring vitactions to individual needs through behavioral profiles and incorporating them creatively into routines, individuals can achieve a deeper, more sustainable sense of balance and fulfillment. This book explores how vitactions can be adapted for specific life stages and applied as non-pharmacological interventions, with particular emphasis on aging, age-related diseases, and cognitive resilience. Notably, childhood and adolescence are pivotal periods during which the presence or absence of vitactions can shape future trajectories of health, resilience, and life outcomes.

Chapter 5. Vitactin Acquisition in Childhood and Adolescence

Childhood and adolescence are pivotal stages of human development, marked by rapid physical growth, cognitive expansion, and emotional maturation. These formative years lay the foundation for a plentiful life, making them critical periods for establishing and nurturing the patterns of action and connection that sustain resilience, adaptability, and creativity throughout life.

Vitactions during childhood and adolescence include playful exploration, curiosity-driven learning, and meaningful social interactions. These actions are not only intrinsic to youthful energy and imagination but also serve as building blocks for emotional intelligence, problem-solving, and cognitive flexibility. Their benefits are far-reaching, influencing academic success, social harmony, and mental health.

Parents, caregivers, educators, and the surrounding environment play a crucial role in fostering these vital activities. By offering opportunities, guidance, and encouragement, they help children integrate vitactions into their daily lives, laying a strong foundation for future challenges. This chapter explores play, learning, and social relationships as foundational vitactions/vitactins, highlights the importance of supportive environments, and examines the opportunities and challenges presented by technology.

5.1 Play and Exploration

Play is a universal and instinctive activity, deeply embedded in human development. For children, it is more than mere recreation, it is a vital process through which they explore themselves, others, and the world around them. Play nurtures curiosity, adaptability, and problem-solving, equipping children to navigate a complex and ever-changing environment.

5.2 Types of Play and Their Benefits

Physical activities, whether running, climbing, or jumping develop motor skills, improve coordination, and enhance overall physical health. These actions release endorphins, which elevate mood and energy levels, and set the stage for habits of movement that contribute positively throughout our lives. A game of tag not only builds stamina but also teaches strategic thinking as children navigate spaces and interactions.

Pretend play, for instance role-playing and storytelling, stimulates creativity and emotional expression. It allows children to experiment with different roles and perspectives, which helps them develop empathy and a deeper understanding of others. An example of imaginative play is a child pretending to be a teacher or doctor, which allows exploration of problem-solving and social dynamics within a safe and playful context.

Group games, collaborative activities, and shared adventures build teamwork, communication, and emotional intelligence. Social play fosters trust, encourages negotiation skills, and strengthens bonds with peers. Building a fort together or playing team sports teaches children how to cooperate, resolve conflicts, and celebrate collective achievements.

Playful exploration encourages risk-taking and experimentation, both of which are vital for developing resilience. The experience of failing, adapting, and trying again within a safe setting prepares children for the more challenging and less forgiving situations they will encounter later in life.

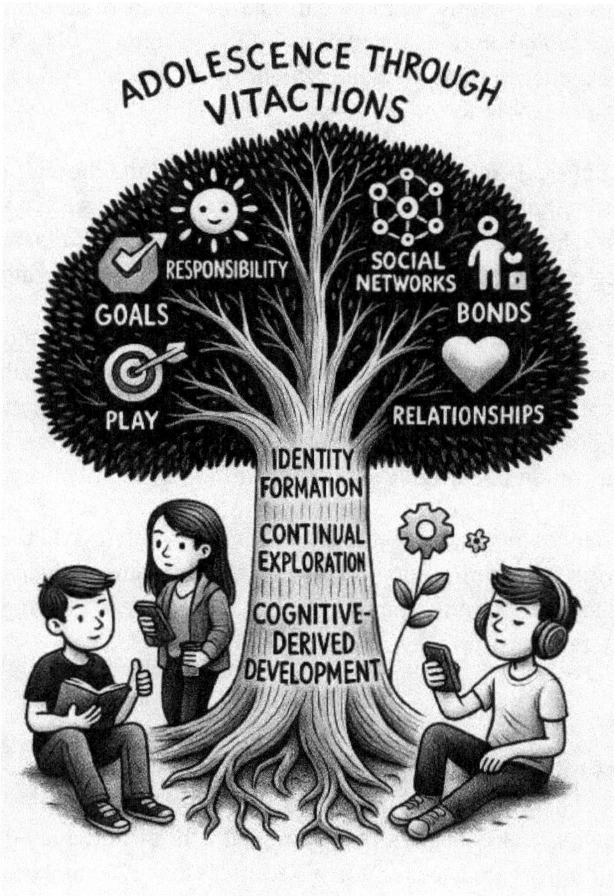

5.3 Learning and Skill Acquisition

The innate curiosity of children and adolescents drives them to explore and learn. This period is marked by a hunger for knowledge and experiences that build resilience and cognitive flexibility. Learning-based vitactions encourage the development of skills that not only provide immediate gratification but also equip individuals with tools for long-term growth.

5.3.1 Examples of Learning-Based Vitactions

1. Music and Arts: Activities like learning to play an instrument, singing in a choir, or engaging in music-based games are among the most versatile vitactions. They activate multiple dimensions, cognitive, emotional, and social, also strengthening memory, coordination, and collaboration. Whether a child is learning to play the piano, dance, or paint, these experiences nurture focus, patience, and fine motor skills and fosters the joy of creative self-expression.

2. Sports and Physical Activities: Mastering physical skills through sports or dance improves coordination, builds self-discipline, and fosters self-confidence. These activities also teach teamwork and perseverance. A teenager participating in basketball develops strategic thinking, physical stamina, and a sense of belonging to a team.

3. Creative Exploration: Activities like drawing, writing, or designing encourage problem-solving and self-expression. They hone critical thinking skills and provide a safe space for children to process their emotions. Encouraging children to write a short story, even if it does not make sense from an adult perspective, fosters imagination and leads to the acquisition of narrative and linguistic skills.

Learning experiences that incorporate play and curiosity often yield the most profound benefits. Experimenting with colors and textures through painting helps a child develop sensory awareness and creative thinking at the same time. Encouraging children to pursue their interests ensures they stay motivated and invested.

5.4 Key Roles of Parents and Caregivers

Parents, caregivers, and educators play a pivotal role in introducing and sustaining vitactions. Their influence shapes children's attitudes toward exploration, learning, and social connection. A supportive environment amplifies the benefits of vitactions, fostering confidence and enthusiasm.

Parents who actively participate in play, learning, or social activities serve as powerful role models. When children see adults valuing these forms of connection, they are more likely to adopt them as essential parts of life. That's why it's a good idea for parents to participate in a family game night to demonstrate the importance of social bonding and shared enjoyment, something that was more natural in ancient times.

Providing opportunities for unstructured exploration allows children to discover their interests and talents. Parents can guide this exploration by introducing new activities and offering gentle encouragement. Enrolling a child in a beginner's art class can spark a lifelong love for creative expression.

Structured activities and freedom should be balanced. Establishing routines ensures consistency, and leaving room for spontaneity nurtures creativity. This balance helps children take part in vitactions without feeling constrained. It is advisable to set aside specific times for playing outdoors or reading as a family, yet still allow for unstructured free time.

5.5 The Role of Environment

Vitactions must provide a sense of safety and accessibility, i.e., they can build up confidence. Environments that are physically and emotionally safe encourage children to take risks and experiment without fear of failure. Parks, libraries, and community centers provide ideal spaces for exploration.

Exposure to varied experiences broadens a child's horizons and increases their adaptability. Access to sports facilities, creative spaces, and cultural events expands the range of vitactions available.

Parents and caregivers also act as gatekeepers, helping children navigate the two most significant challenges, screen time and social pressures, that can interfere with meaningful activities.

5.6 Technology: Opportunity and Threat

Technology has become an integral part of childhood and adolescence, offering both opportunities to enhance vitactions and challenges that may interfere with them. Striking the right balance is key to ensuring that technology supports, rather than displaces, essential forms of connectivity. The rapid pace of technological development is something new that did not occur so rapidly in ancient times; however, ancestral vitactions cannot be replaced by vitactions based on novel technologies. Time will tell what the right balance is.

5.6.1 Opportunities Provided by Technology

Probably the most evident opportunity is the access to Interactive Learning Tools. Apps and platforms designed for educational purposes can transform screen time into active participation. It is important, however, that these apps are appropriate and tailored to children's age and developmental needs. An app that teaches rhythm and melody and allows children to create their own compositions combines learning with creativity.

Collaboration can now be achieved virtually. Digital tools enable group activities and social connections, even across distances. Virtual music lessons or collaborative storytelling platforms allow for shared experiences. A child participating in an online choir learns to harmonize and collaborate, thereby connecting with peers worldwide. Online multiplayer games are also a good option, only if parents set a time limit so that the activity does not become addictive for the child.

5.6.2 Risks of Overuse

Technology-type vitactins must not displace active-type vitactins. Excessive screen time should never replace essential activities such as physical play, imaginative exploration, or face-to-face social interaction. There is also the risk of passive consumption. Watching videos or scrolling through feeds lacks the stimulation and growth opportunities provided by active-type vitactins.

Parents and educators play a crucial role in guiding children's use of technology. By emphasizing interactive and creative tools and, at the same time, setting boundaries around passive consumption, they can ensure technology enhances rather than hinders vitactions.

5.7 Conclusion

Childhood and adolescence are critical periods for embedding vitactions that support lifelong development. Through play, exploration, learning, and social connection, children cultivate the emotional, cognitive, and interpersonal capacities essential for resilience and adaptability. Parents, caregivers, and the environments they create play a pivotal role in nurturing these vital activities, ensuring they become an integral part of a child's life.

Undoubtedly technology offers exciting opportunities to enhance vitactions; however, it must be balanced with active, hands-on forms of participation to unlock its full potential. By integrating vitactions into the daily lives of children and adolescents, we lay the foundation for a lifetime of creativity, connection, and personal growth.

Chapter 6. Adulthood and Aging: The Role of Vitactions in Prolonged Life

Human evolution has left us well-equipped for survival, reproduction, and the challenges of early adulthood, but it did not prepare us for the long lifespans we now enjoy. For most of history, human life expectancy was far shorter than it is today, with individuals rarely living long enough to face the full spectrum of aging-related challenges. Modern advances in medicine, nutrition, and living conditions have extended life expectancy significantly, offering new opportunities for fulfillment but also presenting unique demands on our bodies and minds.

This dissociation between our evolutionary design and the realities of long lifespans creates a need for intentional adaptation. Unlike our ancestors, whose lives were shaped by constant movement, problem-solving, and social bonding, modern adults and seniors often face sedentary routines, cognitive stagnation, and social isolation. Maintaining what supports physical, cognitive, and emotional balance is essential for addressing these contemporary challenges.

This chapter examines the critical role of vitactions in adulthood and aging, exploring their potential to combat chronic diseases, counteract sedentary lifestyles and loneliness, and serve as interventions for aging-related disorders like Alzheimer's and Parkinson's. By understanding and adopting vitactions tailored to the unique needs of these life stages, individuals can navigate the complexities of prolonged life with vitality and good health.

6.1 The Role of Vitactions in Combating Chronic Diseases

Chronic diseases are among the most significant challenges faced during adulthood and aging, often resulting from a combination of genetic predisposition and lifestyle factors. Conditions like cardiovascular diseases, diabetes, and mental health disorders become increasingly prevalent with age, placing physical and emotional strain on individuals and their families. Vitactions offer a proactive and practical approach to managing and even preventing these chronic conditions. In other words, many chronic diseases commonly attributed to aging are strongly influenced by lifestyle and behavioral patterns.

Engaging in physical-type vitactins is one of the most effective strategies for addressing chronic diseases. Incorporating a daily 30-minute walk into one's routine not only lowers the risk of heart disease but also releases endorphins, which alleviate stress and boost mood. Strength-training exercises are equally important, as they help preserve muscle mass, enhance bone density, and support mobility, key factors in preventing frailty and maintaining independence in older adulthood. These activities also contribute to healthy aging and enhance patients' overall well-being. The determination of older individuals training for marathons, or of patients overcoming physical limitations after car accidents or debilitating illnesses, e.g. multiple sclerosis or amyotrophic lateral sclerosis, exemplifies the profound impact of physical activities. Their stories are powerful testaments to the transformative power of movement, inspiring others to push boundaries and embrace an active lifestyle.

Cognition-type vitactions also play a significant role in managing chronic conditions. Activities like solving puzzles, learning a new skill, or engaging in strategy games stimulate the brain, building cognitive reserves that delay memory decline and enhance mental adaptability. They may also contribute to self-satisfaction by offering a creative outlet for emotional experience. Indeed, such activities can be particularly beneficial for individuals with early signs of cognitive impairment, as they promote cognitive arousal and foster a sense of accomplishment.

Practicing mindfulness or engaging in music-based relaxation, are emotional vitactions that provide significant mental health benefits. Chronic stress, often a contributing factor to mental health disorders, can be alleviated through mindfulness practices like deep breathing or meditation. Similarly, listening to calming music during moments of relaxation helps reduce cortisol levels, fostering a sense of peace and emotional stability.

By integrating physical, cognitive, and emotional vitactions into daily routines, individuals can create a holistic approach to managing chronic diseases. This proactive commitment not only mitigates risks but also enhances overall quality of life, allowing adults and seniors to face aging with confidence and vitality.

6.2 Sedentary Lifestyles and Social Isolation

One of the most pressing challenges of modern aging is the prevalence of sedentary lifestyles and social isolation. As adults transition into later stages of life, they often experience a reduction in physical activity and a narrowing of social networks. These changes, whether due to retirement, health limitations, or other factors, have far-reaching consequences for both physical and mental health.

The dangers of a sedentary lifestyle cannot be overstated. Physical inactivity is associated with reduced muscle strength, slower metabolism, and an increased risk of chronic conditions like obesity and diabetes. Furthermore, prolonged periods of sitting contribute to cognitive decline, as reduced blood circulation impacts brain health. An individual who spends most of their day seated in front of a television or computer may experience a gradual decline in both physical stamina and mental clarity. Regular movement is essential to counteracting these effects. Simple yet effective activities like gardening, dancing, or participating in a walking group, encourage physical action and also provide moments of relaxation and enjoyment. A local gardening club not only offers the therapeutic benefits of caring for plants but also fosters social connections and a sense of purpose.

Social isolation, another critical issue, poses risks that extend beyond emotional distress. Loneliness has been linked to increased levels of chronic inflammation, weakened immune responses, and a heightened risk of depression and dementia. Without meaningful interactions, individuals often lack the emotional and cognitive stimulation needed to maintain resilience and adaptability. Vitactions that foster social interactions are essential for combating these effects. Joining a choir or attending community events provides opportunities for emotional connection and shared purpose. Playing an instrument or singing in a group not only strengthens social bonds but also improves mood, creating a sense of belonging and joy.

Volunteering and participating in clubs centered around shared interests, further expand social networks and allow individuals to participate in fulfilling activities. These vitactions remind us that human beings are inherently social creatures, and maintaining meaningful connections is essential for emotional stability and mental health.

6.3 Behavioral Interventions for Aging-Related Disorders

As life expectancy continues to rise, so does the prevalence of aging-related conditions. Particularly concerning is the steady increase in neurodegenerative disorders. These disorders bring unique challenges, profoundly affecting both individuals and their caregivers. Whereas medical treatments provide essential support, vitactions serve as complementary interventions that address the behavioral and emotional dimensions of these conditions.

Physical activity is a cornerstone of managing aging-related disorders. Regular movement promotes neuroplasticity, enhances brain circulation, and reduces systemic inflammation, all of which contribute to maintaining cognitive function. Moderate aerobic exercises like brisk walking or dancing have been shown to slow cognitive decline in individuals with mild cognitive impairment. These activities not only stimulate the brain but also enhance coordination and mobility, supporting multiple aspects of physical and mental vitality. Evidence often overlooked by practitioners shows that patients with Parkinson's disease can exhibit enhanced motor performance and may even appear asymptomatic, when faced with specific challenges, such as stepping up or down curbs[1]. Although seemingly counterintuitive, this strategy enhances goal-directed walking, bypassing the habitual, automatic gait patterns typically impaired in these patients. This distinction is linked to two circuits within the motor control pathways in the brain. Goal-directed walking relies on preserved striatal regions, where proper motor control is compromised due to dopamine depletion in the posterior putamen, underscoring the need to reinforce intentional motor strategies.

Music-based vitactions are particularly effective in managing conditions derived from aging-related neurodegenerative disorders. Singing familiar songs or playing musical instruments stimulates memory and emotional regulation, providing comfort and connection. Music therapy sessions, which awaken both cognitive and emotional systems, offer a unique combination of stimulation and relaxation. Even a person with Alzheimer's may recall cherished memories through music, creating moments of

[1] P. Redgrave *et al.*, "Goal-directed and habitual control in the basal ganglia: implications for Parkinson's disease," *Nat. Rev. Neurosci.*, vol. 11, no. 11, pp. 760–772, Nov. 2010, doi: 10.1038/NRN2915.

clarity and joy that enhance their quality of life. A similar vitaction would be to recall memories from family videos or from recording the voices of family members or close friends.

Cognitive stimulation remains vital in addressing aging-related disorders. Learning a new language, solving complex puzzles, or participating in intellectually engaging discussions delay cognitive decline and improve mental adaptability. Joining a book club that focuses on discussing thought-provoking material sharpens cognitive skills and fosters social connections.

By combining physical, cognitive, and social vitactions, individuals and caregivers can adopt a synergistic approach to managing age-related conditions. A group dance class brings the body and mind into motion and strengthens social connection—offering a holistic path to sustained vitality and quality of life.

6.4 Integrating Vitactions into Daily Life

To reap the full benefits of vitactions, individuals must incorporate them thoughtfully into their daily lives. Establishing consistent routines that incorporate physical, cognitive, and emotional activities helps transform them into lasting habits. Dedicating time each day to a brisk walk, a creative project, and meaningful social interaction fosters a balanced and enriching lifestyle.

New technologies offer additional opportunities for connectivity. Apps and online platforms provide guided exercises, virtual communities, and interactive games, making it easier for individuals to participate in vitactions. A mobile app that tracks physical activity and connects users to virtual walking groups, for instance, promotes both movement and social interaction.

Customization is also essential. Tailoring vitactions to individual preferences, physical abilities, and goals increases their accessibility and effectiveness. An older adult with limited mobility might focus on seated exercises, mindfulness practices, and online social interactions, ensuring they stay active and remain mindful of their capabilities.

6.5 Conclusion

Longer life expectancy brings new challenges, requiring adaptive strategies for living fully and meaningfully. Humans did not evolve to live as long as we do today, making vitactions indispensable tools for navigating the complexities of aging. By addressing chronic diseases, counteracting sedentary lifestyles and social isolation, and providing behavioral interventions for aging-related disorders, vitactions empower individuals to age with resilience, vitality, a sense of purpose, and, above all, health.

As society continues to grapple with the implications of extended lifespans, integrating vitactions into daily routines becomes not just beneficial but essential. These vital activities bridge the gap between our evolutionary design and modern realities, helping ensure that the additional years granted by advances in healthcare and technology are lived with fulfillment and connection.

Table 6.1 presents key recommended activities for each stage of human life.

Life Stage	Suggested Vitactins	Main Objectives
Childhood	Play, storytelling	Cognitive and emotional development
Adolescence	Team sports, art projects	Identity formation, social skills
Adults	Fitness routines, creative activities	Stress management, cognitive stimulation
Elderly	Gardening, choir singing	Social connection, preservation of memory

Table 6.1. *Specific recommendations for each stage of life*

Chapter 7. Vitactins in the 21st Century

Modern society has transformed the way humans live, work, and connect. Advances in urbanization, technology, and convenience have simplified many aspects of life, but have also disrupted essential patterns of living that were once naturally woven into daily routines. The shift from active lifestyles to sedentary habits, combined with increasing reliance on technology, has introduced challenges that undermine physical health, emotional balance, and social connection.

For older adults, these challenges are particularly pronounced. Technology offers opportunities to stay informed and connected, but it also introduces barriers that can create stress and isolation. Whereas digital tools enable global connectivity, they often replace in-person interactions with superficial online connections. Older individuals face additional hurdles; they have difficulties in managing numerous passwords, navigating complex digital platforms, and adapting to the increasing reliance on online services like banking and healthcare. This disconnect between societal advancements and individual needs underscores the importance of reclaiming and adapting vitactions for modern life.

This chapter examines how societal changes, particularly urbanization and technological advancement, have shaped the perception of vitactions. It also explores strategies for reintegrating these essential activities, highlighting their importance in supporting vitality, connection, and well-being, especially among older adults.

7.1 The Loss of Vitactions in Contemporary Society

In traditional societies, vitactions were embedded in daily life, essential for both survival and social cohesion. Activities such as hunting, farming, and gathering demanded constant physical activity, and communal living fostered deep social bonds. Cognitive stimulation arose organically through problem-solving, storytelling, and adapting to environmental changes. Since modern life has disrupted these natural patterns, a new form of vitaction has emerged: caring for domestic animals that require outdoor activity. This not only offers emotional companionship at home but also encourages regular time in nature and spontaneous social interactions with other pet owners, restoring, in part, the movement, connection, and meaning that earlier lifestyles once provided.

Urbanization, with its dense cities and built environments, has limited access to natural spaces. For many, daily life now revolves around desk-bound jobs, long commutes, and sedentary activities at home. These changes have led to reduced opportunities for physical activity, social interaction, and creative exploration. The conveniences of modern technology, online shopping, food delivery, virtual meetings, etc., discourage physical movement and reduce opportunities for interpersonal interactions, contributing to widespread vitaction deficiencies.

For older adults, this shift can be particularly detrimental. Retirement often reduces structured social and physical activities, and urban environments and mobility challenges can lead to isolation. Moreover, the fast pace of urban life leaves little room for the traditional communal practices that once nurtured emotional connection and social cohesion. Reversing these trends requires intentional efforts to reintroduce vitactions into daily routine.

7.2 The Double-Edged Role of Technology

Technology is a defining feature of contemporary society, reshaping how people communicate, work, and interact with the world. For older adults, however, it presents both opportunities and challenges. Its role in accessing vitactions is complex, holding the potential to either enrich or disrupt emotional, cognitive, and social vitality.

7.2.1 The Benefits of Technology

For older adults, technology provides a powerful means of staying connected to the wider world. Video calls, social media, and messaging apps allow individuals to maintain relationships with family and friends, even across great distances. These tools can reduce feelings of loneliness and enable active participation in family events and celebrations.

Technology also opens doors to lifelong learning and hobbies. Apps and online platforms provide access to virtual classes, hobby groups, and cognitive training programs. An older adult interested in painting can join an online art community, sharing their work and receiving feedback from peers around the globe. Similarly, virtual exercise programs or meditation apps enable physical and emotional balance from the comfort of home.

Another significant benefit is the accessibility of essential services through technology. Telemedicine allows older adults to consult with healthcare professionals without the need for travel, and online grocery services ensure that basic needs are met even when mobility is limited. These innovations empower older individuals to maintain independence and access resources that enhance their quality of life.

7.2.2 The Challenges of Technology

Despite its advantages, technology also creates barriers that can undermine its potential benefits. For older adults, adapting to the digital world often comes with significant stress. Managing passwords for multiple accounts, navigating unfamiliar interfaces, and troubleshooting technical issues can be overwhelming. The need to regularly update passwords or remember complex login credentials for banking or healthcare platforms creates frustration and anxiety.

Moreover, while technology fosters global connectivity, it often reduces face-to-face interactions with local friends and family. Virtual conversations, though convenient, lack the depth and emotional resonance of in-person meetings. An older adult may connect with relatives through video calls yet feel disconnected from their immediate community or neighbors.

Another challenge is the design of many digital platforms, which often exclude older users. Complex layouts, frequent updates, and security measures like two-factor authentication can make digital participation daunting. This digital divide not only limits access to the benefits of technology but can also lead to feelings of exclusion and inadequacy.

To fully harness the potential of technology, these barriers must be addressed. Simplifying digital interfaces, offering accessible tutorials, and creating support systems for older users are crucial steps toward ensuring that technology enhances, rather than replaces, vital human connections.

7.3 Cultural and Societal Factors Influencing Vitaction Deficiencies

The modern emphasis on productivity and efficiency has deprioritized essential habits. Social norms often equate busyness with success, leaving little time for reflective or engaging activities. Economic pressures, particularly in urban areas, further limit access to resources, whether fitness programs, creative workshops, or community events, that support vitactions. Even scientists involved in today's fast-paced research, which often produces a flood of publications with minimal genuine advancement in knowledge, find themselves without time to relax and think deeply. Without dedicated time for reflection, achieving groundbreaking discoveries, like those made up to the 1960s, becomes nearly impossible.

For older adults, societal attitudes toward aging can exacerbate these issues. Cultural stigmas often frame aging as a period of decline rather than growth, discouraging older individuals from pursuing new opportunities. Even university professors with active projects are often forced to retirement in many European Universities. This perception creates psychological barriers that prevent involvement in activities like learning a new skill, joining a social group, or embracing technology.

Addressing these cultural and societal influences requires a shift in mindset. Communities must recognize aging as a stage of potential and provide opportunities that empower older adults to reclaim lost vitactions.

7.4 Strategies for Reintroducing Vitactions in Modern Life

Reintroducing vitactions into daily life requires a collaborative effort between individuals, communities, and policymakers. For older adults, addressing technological barriers and creating supportive environments are key strategies.

7.4.1 Designing Spaces for Engagement

Public spaces that encourage movement, interaction, and creativity are essential. Walkable neighborhoods, accessible parks, and community centers provide opportunities for physical activity and social bonding. A park designed with walking paths, fitness equipment, and seating areas for conversations can serve as a hub for older adults to gather, exercise, and connect.

7.4.2 Thoughtful use of Technology

Technology can be challenging; used intentionally, it can also enhance vitactions. Apps designed with older adults in mind, featuring user-friendly interfaces and clear instructions, can support physical, cognitive, and social engagement. An app that offers simple guided exercises or connects users to virtual fitness groups can encourage regular activity.

Online communities also create opportunities for connection and learning. Virtual book clubs, hobby groups, or music workshops allow older adults to build relationships and pursue interests without leaving home. However, these digital interactions should complement, rather than replace, real-world activities, maintaining a healthy balance between virtual and in-person connections.

7.4.3 Addressing Everyday Barriers

Practical solutions can alleviate frustrations with technology. Password managers may simplify the process of managing multiple accounts, and online banking services with dedicated customer support make digital transactions more accessible. Providing in-person tech assistance at community centers or libraries ensures that older adults have the resources they need to navigate digital platforms confidently.

7.5 Behavioral Enrichment in Daily Life

For individuals seeking to reclaim vitactions, small, intentional changes can make a significant difference. Incorporating short walks, attending local events, or dedicating time to hobbies are practical ways to build involvement. Intergenerational programs, where older adults mentor or collaborate with younger participants, foster meaningful connections and enrich both groups.

Mindful use of technology is equally important. Limiting passive screen time and prioritizing active habits helps ensure that technology enriches, rather than diminishes, our cognitive, emotional, and social vitality.

7.6 Conclusion

Modern society offers unprecedented opportunities for connection and growth, but it also disrupts the natural rhythms of physical, cognitive, and social connection. Technology exemplifies this duality, providing tools for global connectivity and personal growth yet also introducing barriers that hinder traditional vitactions. For older adults, navigating this complexity requires thoughtful adaptation and support.

By shaping supportive environments, promoting thoughtful use of technology, and removing barriers to participation, individuals and communities can successfully reintroduce vitactions into daily life. As we move forward, we outline practical, evidence-informed strategies to help people stay active, connected, and fulfilled in an ever-evolving world.

Chapter 8. Vitamins and Vitactions: Parallel Pathways of Health

Vitactions/Vitactins, defined as essential actions that sustain emotional, cognitive, and social health, represent a significant advancement in our understanding of human functioning. Just as vitamins transformed our approach to physical health, vitactins provide a framework for addressing the challenges of modern life by emphasizing practices that nurture mental clarity and social connection.

Like vitamins—well established as essential nutrients for proper bodily function—vitactins play an analogous role in maintaining the health of the mind and behavioral systems. This chapter explores the parallels between vitamins and vitactins, emphasizing their complementary roles. It does not present a historical parallel, as the concept of vitactions is recent. Instead, it builds on the recognition of their importance to draw an analogy: just as vitamins are essential for the physiological health of the body, vitactins are essential for the cognitive, emotional, and social health of the individual.

This analogy helps us understand how behavioral deficiencies mirror physical deficiencies caused by a lack of vitamins.

8.1 Vitamins versus Vitactins

Vitamins and vitactins, though serving distinct domains of health, share a fundamental similarity: both are essential for sustaining complex systems. Vitamins act as the foundational nutrients for the body, enabling cellular processes, supporting immunity, and regulating energy metabolism. Similarly, vitactions/vitactins serve as *behavioral nutrients*, ensuring the smooth functioning of the mind, emotions, and social connections.

Let us exemplify the vitamin-vitactin analogy:

> Vitamin A is critical for vision and immune function, just as social connection is vital for emotional stability and mental resilience.
>
> Vitamin C supports tissue repair and boosts immunity, paralleling the restorative effects of physical movement on both body and mind.
>
> B vitamins play a key role in energy metabolism, akin to how cognitive challenges fuel mental agility and adaptability.

Without these essential inputs, deficiencies arise, leading to systemic dysfunction. The absence of vitamin C results in scurvy, characterized by gum disease, fatigue, and impaired wound healing. Similarly, a lack of social connection can lead to emotional isolation, loneliness, and depression. These correspondences underscore the critical need for both vitamins and vitactins to sustain holistic health.

8.2 The Analogy's Broader Implications

The analogy between vitamins and vitactins becomes clearer when examining specific practices and their roles in maintaining physiological and psychological balance.

Vitactions address deficiencies that are behavioral in nature, mirroring how vitamins address nutritional deficiencies.

Viewing vitactions as behavioral nutrients offers a transformative perspective. This analogy provides a framework for understanding, preventing, and addressing behavioral deficiencies, much as the discovery of vitamins did for nutritional science.

8.2.1 Preventing Behavioral Deficiencies

One of the most profound implications of the vitamin-vitactin analogy is the opportunity to prevent behavioral deficiencies through intentional actions. Just as dietary guidelines and supplementation have been developed to prevent vitamin deficiencies, public health initiatives can promote vitactions to address behavioral deficits. Community programs that encourage social interaction, physical activity, and cognitive enrichment can act as "behavioral supplements" for individuals experiencing isolation, inactivity, or stagnation.

8.2.2 A New Paradigm for Behavioral Health

Viewing daily actions as nutrients offers a powerful way to understand the importance of balance in everyday life. Excessive screen time or overindulgence in certain activities can lead to a state of *hypervitactinosis*, a concept explored in the next chapter. By framing behavioral deficiencies in terms of *behavioral scurvy* or *behavioral rickets*, individuals can better understand the critical importance of maintaining balanced and intentional actions.

8.3 Conclusion

The analogy between vitamins and vitactions/vitactins offers a powerful frame of reference for understanding health as a holistic interplay between body and mind. Like vitamins, vitactions are essential inputs that sustain resilience, adaptability, and overall vitality. The concepts of *behavioral scurvy* and *behavioral rickets*, introduced by the author, highlight the essential role of vitactins in everyday life.

Treating vitactions as behavioral nutrients makes it feasible to develop strategies for integrating these essential habits into daily life—strategies that not only address the challenges of modern living, but also complement nutritional and pharmacological approaches, offering a more comprehensive view of health. The next chapter explores how vitactin excesses and deficiencies impact daily life, emphasizing the crucial role of balance in maintaining health.

Chapter 9. Balancing Behaviors and Nutrition: Excesses and Deficiencies

Health depends on balance — a principle that applies not only to the physical body but also to the rhythms of our daily lives. Just as vitamins must be consumed in appropriate amounts, with deficiencies historically leading to conditions like scurvy or rickets, and excesses causing hypervitaminosis, behavioral inputs also require equilibrium. In the same way, the "intake" of vitactins requires careful balance. *Avitactinosis* disrupts health, leading to emotional isolation, physical stagnation, or cognitive decline. Conversely, excessive involvement in certain habits—a condition proposed here as *hypervitactinosis*—can cause stress, burnout, and other unintended consequences.

The twin challenges of deficiency and excess are here explored, drawing parallels between the physiological effects of vitamin imbalances and the consequences of behavioral imbalances. Both realms demonstrate that health is not just about doing more but about doing enough of the right things in the right proportions. Striking this balance is vital, particularly in today's world, where modern lifestyles often foster behavioral deficiencies or push individuals toward overindulgence. Understanding and addressing these imbalances is essential for maintaining mental and physical vitality and preventing long-term harm.

9.1 Vitactions as Behavioral Nutrients

Avitactinosis arises when essential actions are neglected, much like vitamin deficiencies occur when nutrients are absent from the diet. The absence of certain vitactions leads to conditions that compromise emotional, cognitive, and physical health. These deficiencies are conceptualized as *behavioral scurvy*, *behavioral rickets*, and *behavioral stagnation*, offering a conceptual lens for understanding their impacts.

9.1.1 Social Isolation and Behavioral Scurvy

Social connection is a cornerstone in our lives as human beings, playing a role as fundamental as vitamin C in maintaining immunity and tissue repair. *Behavioral scurvy*, a term proposed by the author, describes the effects of insufficient social life. Individuals experiencing this deficiency often face loneliness, depression, and a diminished ability to handle stress. *Behavioral scurvy* is particularly common among older adults, who may live alone or lack opportunities for meaningful interaction. The absence of regular social connection deprives individuals of the emotional stability and cognitive stimulation that relationships provide, increasing the risk of long-term mental health challenges.

A growing body of converging evidence shows that prolonged social isolation is linked to elevated stress hormones, increased neuroinflammation, and even cognitive decline. The causes of isolation may differ, from mobility limitations to geographic distance from loved ones, but the consequences are strikingly consistent: a deep sense of disconnection that erodes resilience and impairs our capacity to adapt.

9.1.2 Physical Inactivity and Behavioral Rickets

Physical movement is another critical vitaction that the author considers analogous to vitamin D's role in promoting bone health. *Behavioral rickets*, a proposed term, encapsulates the consequences of a sedentary lifestyle. Without regular movement, individuals face a host of negative outcomes, including reduced mobility, weakened muscles, osteoporosis and an increased risk of chronic diseases, mainly obesity, diabetes, and cardiovascular conditions. In behavioral terms, physical inactivity erodes both the body's physical resilience and the mind's capacity to regulate emotions effectively.

Modern life fosters this deficiency in myriad ways. Desk-bound jobs, excessive screen time, and reliance on cars for transportation all contribute to physical inactivity. Streaming television shows or scrolling through social media, which are common leisure activities, often replace opportunities for movement. Over time, the lack of physical activity creates a compounding effect, as diminished strength and energy further discourage participation in physical activities.

9.1.3 Cognitive Stagnation and Behavioral Stagnation

Cognitive activity fosters mental flexibility, adaptability, and creativity, much like B vitamins fuel the body's energy production. *Behavioral stagnation*, another concept proposed by the author, describes the effects of insufficient cognitive stimulation. This condition is characterized by mental fatigue, reduced problem-solving ability, and an increased risk of cognitive decline.

Behavioral stagnation often emerges in environments that fail to challenge curiosity or creativity. Repetitive work tasks or an overreliance on passive entertainment can dull cognitive sharpness over time. Conversely, engaging in activities like learning new skills, solving puzzles, or pursuing artistic endeavors fosters neuroplasticity, enabling the brain to remain agile and adaptable. The absence of such practices creates a void that stifles intellectual growth and diminishes the capacity for innovation.

9.2 Behavioral Excesses: The Rise of *Hypervitactinosis*

Although much attention is given to deficiencies, excessive involvement in certain activities can also disrupt emotional balance, cognitive function, and overall quality of life. Hypervitaminosis, the overconsumption of vitamins, illustrates how even beneficial substances become harmful when taken in excess. Similarly, *hypervitactinosis*, can be used to describe the detrimental effects of behavioral overindulgence. This state arises when individuals perform certain vitactions to such an extent that they undermine the balance required for holistic health.

9.2.1 Overworking and Burnout

In modern society, overworking is one of the most pervasive forms of *hypervitactinosis*. Driven by cultural values that equate productivity with success, many individuals prioritize work at the expense of rest, relationships, and self-care. This excessive focus on professional achievements often leads to burnout, characterized by chronic stress, emotional exhaustion, and reduced motivation. Burnout is not only a psychological condition but also a physical one, as prolonged stress weakens the immune system and increases the risk of chronic illnesses.

A professional who routinely works late into the night may neglect physical activity, skip meals, and withdraw from social connections, triggering a cascade of deficiencies concealed by an overcommitment to work.

9.2.2 Overtraining and Physical Strain

Physical activity is essential for health, but when taken to extremes, it becomes counterproductive. Overtraining, driven by a desire for peak performance or aesthetic ideals, can lead to injuries, hormonal imbalances, and emotional exhaustion. Individuals who exercise compulsively often sacrifice recovery time, leaving their bodies vulnerable to strain and their minds susceptible to anxiety or frustration.

An athlete who trains intensively without adequate rest may experience diminished performance, chronic pain, or even mental burnout. This form of *hypervitactinosis* highlights the importance of balance, even in activities that are fundamentally beneficial.

9.2.3 Excessive Screen Time and Digital Overload

Technology may also contribute to *hypervitactinosis* through excessive screen time. Constant interaction with digital platforms, whether for work, entertainment, or social interaction, results in screen fatigue, reduced face-to-face connections, and mental overload.

Individuals who spend hours scrolling through social media may feel disconnected from their physical surroundings and relationships, despite seeking connection online. The overuse of digital tools creates a paradox where greater connectivity leads to greater isolation, emphasizing the need for mindful use of technology.

9.2.4 Compulsive Cognitive Engagement

Cognitive stimulation is vital for mental health, but too much of it can lead to mental fatigue and burnout. Individuals who constantly seek new intellectual challenges or creative projects without allowing for periods of rest may experience diminished focus, memory lapses, and emotional exhaustion. This form of *hypervitactinosis* illustrates how even positive pursuits can become detrimental when balance is lost.

9.3 Striking the Balance: The Moderation Principle

Achieving balance in vitactions requires intentionality, self-awareness, and adaptability. Just as a balanced diet provides essential nutrients without excess, a thoughtful approach to daily actions helps sustain long-term vitality and mental clarity.

9.3.1 Recognizing Imbalances

The first step in achieving balance is identifying signs of behavioral deficiency or excess. Deficiencies manifest as fatigue, stagnation, or feelings of isolation, and excesses result in burnout, injuries, or mental overload. Regular self-reflection and feedback from trusted individuals can help identify areas that require adjustment.

9.3.2 Cultivating Moderation

Balance involves engaging in a diverse range of activities across physical, cognitive, and social domains, while avoiding over-reliance on any single pursuit. For example, someone who spends much of their time on intellectual work may benefit from adding physical activity and social connection to their daily routine.

9.3.3 Personalizing Vitactions

Balance is not one-size-fits-all. Everyone has unique needs based on their life stage, environment, and goals. A young professional may need strategies to prevent burnout, whereas a retired individual might focus on revitalizing cognitive and social activity.

9.3.4 Embracing Technology with Purpose

Technology can support balance when used intentionally. Apps that track activity, encourage breaks, or provide mindfulness exercises can help maintain harmony. However, it is crucial to limit passive screen time and prioritize practices that foster genuine connection and growth.

9.4 Conclusion

Health, whether physical or mental, depends on balance. Deficiencies in vitactions, like those of vitamins, are harmful, and excessive "intake" creates new challenges. The concepts of *avitactinosis* and of *hypervitactinosis*, introduced in this chapter, provide a basis for understanding how modern lifestyles impact holistic health.

Achieving balance requires intentional effort, self-awareness, and adaptability. By embracing moderation and weaving diverse vitactions into daily life, individuals can cultivate lasting resilience and vitality. As society increasingly acknowledges their importance, future strategies must prioritize the integration of these essential practices into personal routines, community structures, and public health systems.

Chapter 10. Non-Pharmacological Therapies and Vitactions in Aging

Non-pharmacological therapies are increasingly recognized as valuable components in the care of aging populations, particularly in the management of neurodegenerative diseases. These approaches often aim to complement or replace pharmacological interventions, offering patients pathways to improved quality of life through activities that support physical, cognitive, and emotional health. However, a critical flaw in many non-pharmacological therapies lies in their design: they are frequently developed with the convenience of clinicians in mind rather than the natural preferences and routines of patients. Discussions about non-pharmacological interventions at scientific meetings appear, in some cases, to be more focused on generating publications than on genuinely helping patients. Consequently, patients may view these therapies as burdensome obligations, reducing their overall effectiveness. Practitioners and scientists frequently overlook the central principle of personalized medicine: the need to consider each patient as an individual.

In contrast, vitactions offer a more patient-centered approach. Rather than being imposed as rigid tasks, vitactions emphasize activities that are naturally integrated into daily life and aligned with personal interests. They aim to create positive experiences, which makes these activities more sustainable and impactful. This chapter explores the limitations of conventional non-pharmacological therapies, highlights the transformative potential of vitactions, and examines how tools like music and technology can support meaningful participation in later life.

10.1 The Challenges of Conventional Non-Pharmacological Therapies

Conventional non-pharmacological therapies often focus on addressing the symptoms of specific conditions through structured activities. Patients with Parkinson's disease may be encouraged to participate in physical exercises designed to improve mobility and balance, and those with Alzheimer's disease may participate in memory tasks aimed at stimulating cognitive function. Although these interventions can offer significant benefits, they often fail to consider the individual needs, preferences, and emotional readiness of the patient. Furthermore, these rigid approaches often overlook the importance of emotional connection and enjoyment, both of which are essential for long-term adherence.

This clinician-centered model of therapy highlights a broader issue: it assumes that patients will comply with prescribed activities regardless of how they feel about them. However, assumption does not take into account the reality that human beings are more likely to maintain activities that bring them joy or fit naturally into their lives. For therapies to succeed, they must not only address clinical needs but also resonate with the patient's identity and desires.

10.2 Vitactions: A Patient-Centered Alternative

Vitactions represent a paradigm shift in the way we approach non-pharmacological interventions. They emphasize a holistic approach by integrating actions that patients find inherently rewarding. Rather than being imposed, these activities align with the patient's existing routines, preferences, and capabilities, making them easier to adopt and sustain over time.

10.2.1 Physical Vitactions

For individuals with Parkinson's disease, physical vitactions can include dancing, gardening, or walking in a park. These actions integrate movement with joy and purpose, turning exercise into an enjoyable and meaningful part of daily life. Dancing to a favorite song not only enhances coordination and mobility but also elevates mood and fosters social interaction. Similarly, instructing Parkinson's patients to walk following specific patterns, such as repeatedly thinking about stepping up and down a curb (as mentioned above in **6.3**), represents an appropriate and effective action tailored to their needs.

10.2.2 Cognitive Vitactions

Cognitive vitactions can be designed to meet the needs of patients with Alzheimer's disease and other forms of cognitive decline. Rather than relying on rigid memorization tasks, these activities stimulate the mind in more organic ways. Storytelling, creative writing, or exploring new music can provide cognitive challenges without inducing frustration. Listening to familiar songs or familiar voices can also evoke powerful memories, offering moments of clarity and emotional connection.

10.2.3 Social Vitactions

Social interactions are a cornerstone in our lives, particularly for older adults who may face isolation. Group listening sessions or intergenerational storytelling, are vitactions that, by encouraging shared experiences, provide both emotional support and cognitive stimulation. These activities foster a sense of belonging, reducing feelings of loneliness and enhancing overall resilience.

10.3 Music as a Vehicle for Vitaction Expression

Music stands out as a universally engaging and accessible tool. Its ability to stimulate emotional, cognitive, and social fulfillment makes it an ideal intervention for aging individuals, regardless of their physical or cognitive abilities. For millennia, music has served as entertainment, for disease prevention, and for health recovery.

10.3.1 Emotional Benefits

Music has a profound impact on emotions, offering comfort, joy, and relaxation. Listening to favorite songs or calming instrumental music can reduce anxiety, elevate mood, and create a sense of peace. For individuals with neurodegenerative conditions, music often serves as a bridge to positive memories and emotions, fostering moments of happiness even in challenging circumstances.

10.3.2 Cognitive Stimulation

Music lights up multiple areas of the brain, making it a powerful cognitive stimulator. For Alzheimer's patients, songs from their youth can trigger vivid memories and enhance focus. Music-based activities stimulate cognitive function and support neuroplasticity. They can range from recognizing instruments in a piece or recalling lyrics to learning to play a new song on an instrument.

10.3.3 Social Connection

Music can bring people together, creating shared experiences that strengthen bonds. Whether through singing in a group, attending a live performance, or discussing favorite artists, music fosters communication and connection. These interactions are particularly valuable, who may struggle to maintain social networks as they age.

10.4 Practical Strategies for Integrating Vitactions

The success of vitactions depends on their seamless integration into patients' lives. This requires a shift from prescriptive therapy models to approaches that prioritize personalization, flexibility, and enjoyment.

10.4.1 Incorporating Physical Vitactions

Physical vitactions should focus on activities that are both therapeutic and enjoyable. Walking in nature, dancing, or participating in light gardening provide physical benefits and engage the senses, lifting the spirit. These activities can be tailored to the patient's abilities, ensuring that they remain accessible and rewarding.

10.4.2 Introducing Cognitive Vitactions

Cognitive vitactions can be embedded in everyday leisure activities. Listening to music, solving puzzles, or engaging in storytelling sessions offers cognitive stimulation in a relaxed and enjoyable context. Apps like *Radio Garden*[2] further enrich this experience by fostering active exploration and global awareness. Beyond music, the platform enables users to listen to news, talk shows, cultural programs, and local perspectives from around the world, offering opportunities to learn about current affairs, understand social contexts in different countries, and practice foreign languages. By providing access to more than 8,000 radio stations worldwide, *Radio Garden* supports curiosity-driven involvement, intercultural learning, and sustained cognitive stimulation in an accessible and entertaining way.

---

[2] **Radio Garden**: An application and web platform that allows users discover and tune in to a myriad of radio stations worldwide in real time through an interactive map. It offers stations in numerous languages and musical genres. When playing music stations, it displays on screen the song title and the performer—or the composer, in the case of classical music—enhancing the listening experience and cultural discovery.

10.4.3 Encouraging Social Engagement

Social vitactions are essential for combating loneliness and fostering connection. Attending concerts, joining community music sessions, or engaging in intergenerational storytelling provides meaningful opportunities for interaction and emotional support. Virtual platforms can also facilitate connections for patients who are homebound, ensuring that they remain socially connected.

10.4.4 Leveraging Technology

Technology offers countless opportunities to support vitactions. Apps and online platforms can provide guided exercises, mindfulness practices, or virtual group interactions, making it easier for patients to engage in meaningful activities. However, it is essential to ensure that these tools are user-friendly and tailored to the needs of older adults, who may face barriers to digital literacy.

10.5 Conclusion

Non-pharmacological therapies hold great promise for improving quality of life in aging populations, but their effectiveness depends on thoughtful design. Traditional approaches often fall short by prioritizing clinical outcomes over the lived experiences of individuals, resulting in disengagement and frustration. Incorporating appropriate vitactions offers a compelling alternative by emphasizing activities that are personally enjoyable and meaningful, thereby promoting sustained involvement and lasting impact.

By integrating natural activities such as music listening, storytelling, and gentle movement, vitactions transform therapy into a source of joy and connection. Tools like *Radio Garden* illustrate how technology can enrich these experiences, offering moments of relaxation, cognitive stimulation, and social fulfillment. Looking ahead, the focus must shift toward designing therapeutic interventions that patients willingly embrace—so that vitality, connection, and emotional richness become a natural part of daily life.

Chapter 11. Vitactions in Healthcare

Modern healthcare systems often prioritize pharmacological treatments as the default solution for chronic conditions. Medications are undeniably valuable, but their limitations, including side effects, diminishing effectiveness, and failure to address underlying behavioral causes, are well-documented. As the global burden of chronic diseases grows, there is an increasing recognition of the need for approaches that go beyond pills and injections. Essential activities, if naturally integrated into daily life, offer a revolutionary approach to healthcare by complementing or even replacing traditional medical interventions.

Vitactions are more than supplementary tools; they represent a paradigm shift in healthcare. Unlike rigid clinical routines or generic prescriptions, vitactions are patient-centered, prioritizing practices that align with individual preferences and routines. Whether through physical activity, social life, or cognitive stimulation, vitactions can empower patients to take an active role in their health. This chapter explores how vitactions can enhance healthcare outcomes, drawing on case studies that suggest their effectiveness in managing Alzheimer's disease, obesity, and diabetes. It also examines the challenges and opportunities involved in integrating vitactions into mainstream healthcare.

11.1 Behavioral Therapy Through Vitactions

Behavioral therapy has long been recognized as an effective adjunct to traditional medical treatments, addressing the habits and patterns that contribute to chronic diseases. Vitactions take this concept a step further by emphasizing natural, enjoyable activities that patients are more likely to sustain. Unlike prescribed routines that often feel burdensome or disconnected from everyday life, vitactions are seamlessly woven into the fabric of daily living.

Consider a patient managing stress-related hypertension. Traditional advice might focus on medication and a strict exercise regimen, which can feel overwhelming. A vitaction-based approach, on the other hand, might include listening to calming music, taking short nature walks, or practicing mindfulness exercises during moments of stress. They are not only effective but also enjoyable, making them far more sustainable in the long term.

The adaptability of vitactions is one of their greatest strengths. Not only can these activities be tailored to suit a wide range of abilities, preferences, and cultural contexts, but individuals themselves respond in unique ways to different types of vitactions. What energizes or comforts one person may not resonate with another, making personalization key to effectiveness. An older adult with limited mobility might find meaning and benefit in seated yoga or storytelling with family members, and a younger, more active individual may prefer jogging, dancing, or joining group sports. This two-way flexibility—both in how vitactions are designed and in how people take part—ensures their relevance and accessibility across diverse populations.

11.2 Case Studies of Vitactions in Practice

Alzheimer's disease presents complex challenges, including cognitive decline, anxiety, and social isolation. Traditional therapies often focus on pharmacological interventions that may alleviate symptoms but do little to address the broader needs of the patient. Vitactions, however, provide holistic support by engaging the mind, emotions, and social connections.

Music, for example, has emerged as a powerful vitaction for Alzheimer's patients. Listening to familiar songs can trigger vivid memories and emotional responses, creating moments of joy and clarity.

The app *Radio Garden* exemplifies how technology can enhance musical vitactions and broader cognitive engagement.

Individuals across a broad cognitive spectrum, from patients with cognitive impairment to cognitively healthy older adults, can interact with the app in several ways:

> Passive Listening: Tuning into a station that plays soothing or familiar music, spoken-word programs, or news broadcasts for relaxation and enjoyment.
>
> Active Listening: Guessing the name of the song, artist, or composer —or identifying the language, country, or topic being discussed in talk or news programs— as a fun cognitive exercise.
>
> Social Interaction: Discussing musical tastes, cultural differences, current events, or linguistic features with family members or caregivers, creating moments of shared discovery and conversation.

By incorporating both passive and active elements, *Radio Garden* transforms listening (whether to music, voices, or global narratives) into a multi-dimensional vitaction that stimulates the mind, nurtures emotions, and fosters social bonds.

Storytelling is another valuable vitaction for patients with mild cognitive impairment (MCI) that often precedes Alzheimer's disease. Sharing personal narratives or listening to stories from others stimulates memory and strengthens social bonds. These activities not only provide cognitive benefits but also create opportunities for meaningful connection, reducing feelings of isolation.

Diabetes management requires a careful balance of glucose control, physical activity, and stress reduction. Vitactions provide practical tools for achieving this balance, complementing traditional treatments like insulin therapy.

Managing obesity often requires addressing complex behavioral patterns related to diet, activity, and stress. Traditional interventions like calorie counting or restrictive diets can be difficult to maintain, leading to cycles of frustration and relapse. Vitactions, by contrast, focus on sustainable behavioral changes that promote physical and emotional vitality. Mindful eating, another vitaction, encourages patients to pay attention to hunger cues, savor meals, and identify emotional triggers for overeating.

These practices foster a healthier relationship with food, addressing the root causes of weight gain.

Physical vitactins offer enjoyable ways to incorporate movement into the daily life of individuals with obesity. A patient might choose to dance to their favorite music for 15 minutes a day, combining physical activity with emotional uplift. Alternatively, they may opt for a daily walk with a pet or a friend, turning exercise into a moment of connection and routine.

Stress-reducing vitactions also play a crucial role in obesity management. Activities like journaling, listening to calming music, or practicing yoga help mitigate the physiological effects of stress, which is often a major contributor to weight gain. Light physical activities -yoga or stretching- improve insulin sensitivity, and mindfulness exercises help regulate cortisol levels, which can impact blood sugar. By addressing both physical and emotional dimensions, vitactions offer a comprehensive approach to managing obesity.

Creating a relaxing playlist or exploring music through favorite local radio stations or apps, can reduce stress and promote emotional resilience. Social-type vitactins, like joining a walking/jogging group or participating in community activities, provide additional benefits by fostering connection and support. These activities not only improve physical health but also support mental clarity, empowering patients to take an active role in managing their condition.

11.3 The Patient-Centered Model of Vitactions in Personalized Medicine

At the heart of vitactions lies a patient-centered approach to healthcare. Unlike conventional therapies that prioritize clinical convenience, vitactions focus on the individual, tailoring interventions to their unique preferences, abilities, and routines. This shift from prescriptive to personalized care enhances adherence and fosters a sense of empowerment.

A patient who loves music might participate in regular listening sessions or use apps to explore new genres. Someone who enjoys nature might incorporate walking or gardening into their daily routine. Patients who value family connections might participate in storytelling or shared activities with loved ones. By aligning with patients' interests, vitactions transform therapy from an obligation into a source of joy and fulfillment.

11.4 Challenges and Opportunities

Integrating vitactions into healthcare is not without challenges. The current system often favors pharmacological solutions due to their scalability and ease of implementation. Behavioral therapies, by contrast, require personalization and interdisciplinary collaboration, which can be resource intensive. It should be noted that pharmacological interventions for several chronic diseases, such as Alzheimer's disease, Huntington's disease, multiple sclerosis, and amyotrophic lateral sclerosis, remain limited in effectiveness.

To overcome these barriers, healthcare providers must be educated about the value of vitactions and trained to integrate them into care plans. Policymakers and insurers also have a role to play in supporting vitaction-based interventions through funding and reimbursement policies.

Technology offers significant opportunities to expand access to vitactions. It is now easy for patients to begin music-based activities, and wearable devices can monitor physical activity and deliver reminders for mindfulness exercises. By leveraging digital tools, healthcare systems can deliver vitactions at scale, yet still maintain a personalized touch.

11.5 The Future of Vitactions in Healthcare

The integration of vitactions into healthcare represents a promising shift toward more holistic and sustainable care. By addressing the limitations of pharmacological treatments and focusing on actions that align with patients' daily lives, vitactions offer a pathway to greater vitality, resilience, and quality of life. Public health campaigns can amplify this impact by promoting vitactions as essential components of preventive and therapeutic care.

Community programs that encourage walking, dancing, or shared music sessions can attract people who might otherwise feel isolated. Policymakers can advance this shift by funding initiatives that foster vitaction-friendly environments, including the development of green spaces, community centers, and music-based therapy programs. Such efforts help ensure that vitactions become not only individual choices but also shared societal priorities.

11.6 Conclusion

Vitactions have the potential to transform healthcare by providing patient-centered alternatives to traditional treatments. Whether through music, physical activity, or storytelling, vitactions address the physical, cognitive, and emotional dimensions of health in ways that are both effective and enjoyable.

By integrating vitactions into health systems, we can move towards a model of care that is not only therapeutic, but also empowering and satisfying. As this chapter demonstrates, the possibilities for vitactions in healthcare are vast, from managing chronic conditions to enhancing quality of life. The next chapter will explore the research needed to validate and refine vitaction-based interventions, paving the way for their widespread adoption.

Chapter 12. Vitaction-Centered Research

As a novel concept, vitactions require rigorous scientific exploration to understand their full potential, particularly their impact on the brain. Much like the discovery of vitamins spurred the development of nutritional science, the study of vitactions could transform the understanding of how they influence brain health, emotional resilience, and social connectivity. However, to advance this field, it is essential to classify vitactions systematically, enabling their targeted application in health interventions. This chapter explores how vitactions may influence brain structure and function, presents a conceptual structure for their classification, and underscores the need for research to validate their effects and support their integration into healthcare systems.

12.1 The Neurological Impact of Vitactions

Understanding how vitactions influence the brain is essential to establishing their role in human health. Vitactions are not isolated actions; they recruit complex neural networks and activate mechanisms that foster adaptation, growth, and resilience. They impact three core areas of brain function: neuroplasticity, emotional regulation, and social connectivity.

12.1.1 Neuroplasticity

Learning a new language, solving puzzles, or playing a musical instrument stimulates neuroplasticity, the brain's ability to reorganize itself by forming new neural connections. This adaptive capacity is essential for learning, memory, and recovery from injury or disease. Engaging in storytelling or creative writing recruits brain regions involved in language, imagination, and memory retrieval. Over time, these vitactions strengthen neural pathways, enhancing cognitive flexibility and adaptability.

In the case of aging populations, neuroplasticity becomes especially important. Regular involvement in cognitive vitactions may slow age-related decline, preserving memory and decision-making skills. Research has shown that activities requiring sustained mental effort, not only protect against cognitive deterioration but also foster a sense of accomplishment and purpose.

12.1.2 Emotional Regulation

Vitactions also play a crucial role in managing emotions. Many of the above-described activities, either physical, emotional or social, trigger the release of neurotransmitters like dopamine, serotonin, and endorphins. These chemicals regulate mood, reduce anxiety, and promote feelings of happiness. A person who incorporates regular dance sessions into their routine may experience enhanced emotional stability, as the combination of physical activity and rhythmic movement activates brain regions associated with pleasure and reward.

Moreover, self-expressive vitactions, such as painting, provide outlets for processing emotions. These actions help individuals navigate complex feelings, reduce stress, and develop resilience to life's challenges. Over time, such practices create emotional habits that buffer against depression and anxiety.

12.1.3 Social Connectivity

Human beings are inherently social, and group discussions, storytelling, and collaborative projects are vitactions that strengthen connections and activate brain regions involved in empathy, communication, and social bonding. These interactions not only combat loneliness but also stimulate cognitive and emotional functions, enhancing overall mental health.

When individuals share stories or participate in group activities, they activate the prefrontal cortex and limbic system, which are critical for understanding others and forming emotional connections. This activation is particularly valuable for older adults, who may face isolation due to reduced mobility or changing family dynamics. These vitactions help strengthen social ties, fostering both individual satisfaction and community cohesion.

12.2 Further Advances in Vitactin Classification

As research into vitactions progresses, the need for a systematic classification system becomes increasingly apparent. Just as vitamins are categorized based on their roles in physiological processes, vitactions can be physical or grouped according to their primary impacts on mental, emotional, and social health. Such a taxonomy would not only enhance understanding but also guide interventions tailored to specific needs.

12.2.1 Physical Vitactins

Activities like walking, dancing, or yoga fall into this category, as they promote movement, improve cardiovascular health, and enhance metabolic function. Physical vitactions are particularly effective in reducing stress and elevating mood, as they combine physical exertion with mental focus.

12.2.2 Cognitive Vitactins

Problem-solving, acquiring new skills, and engaging in creative pursuits stimulate neuroplasticity and build cognitive resilience. Vitactions that enhance cognition are therefore invaluable for preventing or mitigating the effects of neurodegenerative diseases, especially when practiced consistently over time.

12.2.3 Emotional Vitactins

These actions, including mindfulness, music-based activities, and journaling, regulate mood and build emotional stability. Emotional vitactions help individuals manage stress, navigate complex feelings, and cultivate inner peace.

12.2.4 Social Vitactins

Group activities, collaborative projects, and shared storytelling foster connection and empathy. Social vitactions are essential for mitigating loneliness and promoting a sense of belonging, particularly in vulnerable populations. They are especially valuable for people experiencing loneliness and for individuals recovering from illness.

This classification is not merely academic; it provides a practical roadmap for designing interventions that address specific health challenges, ensuring that vitactins are applied in ways that maximize their impact.

12.3 Curated Databases and Longitudinal Studies

To refine the classification and application of vitactions, researchers must gather comprehensive data on their adherence and effects. Curated databases are essential to this effort, capturing information across diverse populations and settings. They would include metrics on daily activity patterns, health outcomes, and environmental factors—providing a rich foundation for analysis and insight

Tracking the frequency and type of musical activity across age groups may reveal patterns in how this vitaction supports emotional regulation and cognitive health. Likewise, examining the effects of storytelling on memory and social connection in older adults could highlight its potential as a therapeutic intervention.

Longitudinal studies are equally important, as they enable researchers to observe how vitactions influence health trajectories over time. Such studies can reveal critical periods when specific vitactions offer the greatest benefits, such as early childhood for cognitive development or later life for social connection. By tracking individuals across the lifespan, researchers can generate robust evidence of efficacy, supporting the inclusion of vitactions in public health strategies.

12.4 Biomarkers and Predictive Modeling

Advances in technology offer exciting possibilities for measuring and personalizing vitactions. Biomarkers, whether hormonal changes or neural activity patterns, can provide objective indicators associated with how vitactions impact the brain and body. A reduction in cortisol levels following mindfulness practice could serve as a marker of stress reduction, and increased neuronal activity in the prefrontal cortex may underlie improved cognitive performance during problem-solving tasks.

Predictive modeling, powered by artificial intelligence, further enhances the potential of vitactions. By analyzing health history and behavioral preferences, predictive tools can recommend specific vitactions tailored to optimize outcomes for individuals. A personalized vitaction plan for a patient recovering from surgery might emphasize light physical activities and social connections, based on predicted benefits for recovery.

12.5 Interdisciplinary Collaboration

The study of vitactions requires collaboration across multiple disciplines. Neuroscience offers insights into how specific practices influence brain function, whereas psychology helps uncover the motivations and obstacles that shape engagement with vitactions. Public health initiatives may focus on designing strategies to promote vitactions on a societal level, ensuring their accessibility to diverse populations. Technology, too, plays a crucial role, with apps and wearables offering tools to monitor, encourage, and track vitaction "intake".

By bringing these fields together, researchers can develop a comprehensive understanding of vitactions and translate their findings into practical applications. Partnerships between academic institutions, healthcare organizations, and tech companies can accelerate progress, ensuring that the benefits of vitactions reach individuals and communities worldwide.

12.6 Conclusion

The journey to understanding vitactions is akin to the early days of vitamin research, where scientific inquiry transformed scattered observations into a structured field of study. By examining how vitactions influence the brain and by categorizing them by function, researchers can unlock their full potential as tools for strengthening mental, emotional, and social stability. This effort requires curated databases, longitudinal studies, and interdisciplinary collaboration, all aimed at building a robust evidence base for vitactions.

As the science of vitactions evolves, it holds the promise of reshaping behavioral health, offering new pathways to resilience, connection, and fulfillment. The next chapter will focus on practical strategies for integrating vitactions into daily life, informing public health policy development and population-level implementation.

Chapter 13. Integrating Vitactions into Daily Life

The success of vitactions lies not only in their theoretical foundation but also in their practical integration into everyday life. Vitactions are most impactful when they align with daily routines, creating moments of joy, connection, and self-discovery. To achieve this, individuals, communities, and policymakers must work together to design environments, tools, and strategies that support the natural adoption of vitactions.

Unlike classical clinical interventions, which can feel rigid and prescriptive, vitactions thrive on personalization and adaptability. They are not tasks to be completed but enriching experiences to be embraced. For this vision to become a reality, the concept of 'vitaction' and its significance must be communicated widely. Awareness is the first step toward adoption, and tools must be developed to help people understand and integrate vitactions into their lives. This chapter explores how environments, technology, education, and public policy can make vitactions an integral part of human culture, benefiting individuals and society at large.

13.1 Designing Vitactin-Enriched Environments

Environments profoundly influence our way of life. A neighborhood designed for walkability invites residents to incorporate physical vitactions, and a community center offering group activities fosters social connection. Creating spaces that naturally support vitactions is a vital strategy for ensuring their sustained practice.

The physical layout of our surroundings plays a crucial role in encouraging or discouraging specific activities. Imagine a neighborhood with wide sidewalks, green parks, and safe bike paths. Such an environment invites residents to take walks, cycle, or do light exercise, making physical vitactions a natural part of daily life. In contrast, urban areas with heavy traffic, limited green spaces, and poor walkability discourage movement and isolate residents.

Green spaces, in particular, offer opportunities for multiple types of vitactions. Parks and community gardens not only encourage physical activity but also serve as venues for social interaction, mindfulness, and connection with nature. Designing these spaces with accessibility in mind ensures that individuals of all ages and abilities can participate in vitactions.

Well-designed indoor environments can play a key role in supporting vitactions. At home, this might involve setting aside spaces for creativity, relaxation, or physical activity, such as a quiet corner for yoga or a shelf filled with puzzles and games. Likewise, workplaces that prioritize *joie de vivre*—through ergonomic setups, communal spaces, and opportunities for movement throughout the day—make it easier for employees to integrate vitactions into their daily routines.

Social environments are just as important as physical ones in fostering vitactions. Community hubs may be libraries, cultural centers, and recreational facilities; they provide opportunities for storytelling sessions, group music activities, or shared artistic endeavors. These structures not only invite social vitactions but also weave stronger communal ties, nurturing a sense of belonging and shared human richness.

Social structures can also take informal forms. Neighborhood walking groups, book clubs, and community gardens are grassroots initiatives that foster connection and encourage participation in vitactions in a natural and enjoyable way.

13.2 Technology and Vitactions: A Synergy for Well-being

Technology has transformed how we connect, learn, and interact with the world, making it a powerful tool for promoting vitactions. Apps, digital platforms, and wearable devices can guide individuals toward meaningful actions that enhance daily life, offering both structure and inspiration.

One example of how technology supports vitactions is through music streaming platforms that offer access to a vast range of radio stations and playlists from around the world. These tools allow individuals to explore music across genres and cultures, engaging in both passive and active vitactions. Listening to favorite songs can uplift emotions and reduce stress, and identifying a song, artist, or composer stimulates cognitive functions. Sharing musical discoveries with friends or family adds a social dimension, turning a solitary moment into a shared experience.

Mindfulness apps offer another avenue for vitactions. By guiding users through meditation exercises, breathing techniques, or journaling prompts, these tools encourage emotional balance and stress relief. Similarly, fitness trackers that remind users to move or celebrate milestones can motivate physical vitactions, turning abstract health goals into tangible achievements.

Although technology can enhance vitactions, it is essential to avoid overreliance. Excessive screen time or passive consumption of digital content can undermine the very goals of vitactions, leading to isolation or mental fatigue. To strike the right balance, technology should serve as a bridge to real-world connection, not a replacement for it. A virtual storytelling session might connect distant family members, but in-person gatherings should remain the priority when feasible.

13.3 Communicating the Concept of Vitactions

For vitactions to become a universal practice, the word and its significance must be widely communicated. Public awareness is the cornerstone of adoption. People need to understand what vitactions are, why they matter, and how they can enhance their lives. Without this knowledge, even the most well-designed environments and tools will fall short.

13.3.1 Educational Campaigns and Media Outreach

Educational campaigns can introduce vitactions to diverse audiences through schools, workplaces, and community programs. Vitactions can be incorporated into health or psychology curricula, helping children and young adults understand the importance of actions that support emotional balance, cognitive growth, and social connection. Media outreach, including documentaries, podcasts, and social media campaigns, can further spread the message, making vitactions a part of everyday conversation.

13.3.2 Workshops and Community Engagement

Workshops and seminars are another effective way to promote vitactions. A local community center could host sessions on incorporating vitactions into family life. These events not only educate participants but also create immediate opportunities for participation.

13.3.3 Developing Tools to Support Vitactions

Practical tools can make vitactions accessible to everyone. An app designed to suggest personalized vitactions based on an individual's preferences and lifestyle could help users identify and integrate them into their routines. Online guides, videos, and tutorials can provide step-by-step instructions for beginners, and forums and social networks can connect individuals seeking to share their experiences.

13.4 Practical Guidance for Individuals

Incorporating vitactions into daily life starts with small, intentional steps. These actions should feel natural and enjoyable, enhancing existing routines rather than adding pressure or complexity.

Individuals can identify opportunities to incorporate vitactions into their daily schedules. A morning walk combines physical movement with mindfulness and connection to nature. Similarly, family meals can become moments of storytelling or shared music experiences, adding depth and connection to an everyday habit.

Common barriers include lack of time and motivation. These can be addressed initially through small, manageable vitactions that help build confidence and create momentum. Celebrating progress, no matter how small, reinforces positive habits and encourages continued adherence. Involving friends or family adds accountability and makes vitactions more enjoyable.

13.5 Strategies for Communities and Policymakers

Communities and policymakers play a decisive role in creating environments in which vitactions can flourish and become a natural part of everyday life. At the community level, the organization of open activities, such as walking groups, creative workshops, cultural gatherings, or volunteer initiatives, provides genuine opportunities for active participation. These events not only facilitate the practice of vitactions but also strengthen social bonds, foster cooperation, and help build a shared sense of purpose and mutual support. When communities mobilize, opportunities to adopt healthy habits multiply.

From a legislative perspective, public policies can generate profound and lasting impact. The development and maintenance of parks, walking paths, green spaces, sports facilities, and cultural centers create accessible settings in which vitactions can be practiced without economic or social barriers. Investment in community programs, mental health services, and other supportive initiatives further expands these opportunities, particularly for individuals who might otherwise be excluded. Moreover, measures such as promoting flexible working arrangements, encouraging lifelong learning, and supporting local civic projects make vitactions easier to integrate into daily routines.

Taken together, coordinated action by communities and policymakers can reshape the social environment, ensuring that vitactions are not the exception but an accessible and everyday possibility for all.

13.6 Cultural and Lifespan Adaptations

Vitactions are not one-size-fits-all; they must be tailored to cultural contexts and life stages. In childhood, play-based vitactions that encourage exploration and imagination are critical for cognitive and emotional development. For adults, activities that challenge the mind and foster social connections support personal growth and professional success. Older adults benefit from gentle physical activities and social interactions that combat isolation and maintain vitality.

Recognizing and adapting to these variations ensures that vitactions remain relevant and inclusive, meeting the unique needs of individuals across different cultures and phases of life.

13.7 Conclusion

Integrating vitactions into daily life requires a collective effort. Individuals, communities, and policymakers each have a role to play in creating environments and opportunities that make it easier to take part in essential actions. The word "vitactin" and its significance must be communicated widely, ensuring that everyone understands its value and can access tools to incorporate vitactions into their lives.

By embracing vitactions, individuals enrich their lives and help build stronger, more connected communities. As a society, we have the power and the need to make vitactions a natural and enriching part of human culture, fostering resilience, joy, and connection for generations to come.

As modern societies seek to reclaim the health-promoting potential of vitactions, it is worth turning to cultures that have long preserved these practices as pillars of daily life. This brings us to Asian traditions, where vitactions remain deeply woven into the fabric of physical health and mental harmony.

Chapter 14. Vitactions in Asian Traditions: Tools for Body and Brain Health

14.1 Ancestral Traditions of Asian Origin

Throughout history, Asian cultures have cultivated practices that emphasize the harmonious integration of body and mind, recognizing their interdependence in achieving health and resilience. Practices such as yoga, tai chi, and reiki, resonate with the emerging concept of vitactions. Despite originally rooted in spiritual or philosophical traditions, these activities have demonstrated therapeutic potential in modern contexts, especially for chronic diseases like neurodegenerative disorders, obesity, and diabetes. This chapter explores the roles of these practices in maintaining health, restoring balance, and addressing the challenges of modern lifestyles.

14.1.1 Yoga: The Union of Mind and Body

Originating in ancient India, yoga is a comprehensive discipline that encompasses physical postures (asanas), breath control (pranayama), and meditation. Modern research validates yoga's capacity to act as a vitaction by addressing the physiological and psychological imbalances caused by sedentary lifestyles and stress. As mentioned in previous chapters, yoga should not be introduced in an unnatural way; it can be introduced depending on the individual's willingness to experiment with new tools to see if they gain benefits.

Yoga's efficacy lies in its ability to modulate the hypothalamic-pituitary-adrenal (HPA) axis, reduce cortisol levels, and enhance parasympathetic activity. These effects are particularly beneficial for managing chronic stress, a precursor to many lifestyle diseases. Regular yoga practice increases levels of brain-derived neurotrophic factor (BDNF), supporting neuronal health and plasticity. It has shown promise in mitigating cognitive decline associated with Alzheimer's disease and age-related memory loss. Yoga also improves insulin sensitivity and reduces biomarkers of inflammation, making it effective in managing type 2 diabetes and obesity. Additionally, in easing anxiety and depression, yoga cultivates emotional resilience, helping to anchor the mind in clarity and calm.

14.1.2 Tai Chi: Moving Meditation

Tai chi, a Chinese martial art characterized by slow, deliberate movements, integrates physical exercise with mindfulness. It is often described as "meditation in motion" and is adopted for its health-promoting benefits.

Tai chi's rhythmic movements enhance physical and mental health by improving balance, coordination, and cardiovascular fitness. Studies have linked tai chi practice to improved executive function and memory, potentially delaying cognitive impairment in older adults. It improves circulation and reduces blood pressure, helping to mitigate the risk of cardiovascular diseases, often comorbid with diabetes and obesity. Tai chi also improves proprioception and balance, reducing falls and related injuries and promoting longevity.

14.1.3 Reiki: Energy Healing for Restoration

Reiki, a Japanese complementary therapy, is typically delivered through gentle, non-invasive hand placements either on or just above the body. Although the mechanisms underlying the reported benefits are less understood compared to yoga and tai chi, reiki is valued for its calming effects and its ability to complement conventional treatments.

Reiki addresses the mind-body interface by inducing a deep state of relaxation, alleviating chronic pain and stress-related disorders. It has been used alongside conventional therapies for cancer, diabetes, and neurodegenerative diseases, enhancing patients' quality of life. Reiki's emphasis on energy balance supports emotional resilience and mental clarity, contributing to its reputation as a holistic therapy.

14.2 The Vitaction Perspective. Bridging Tradition and Modernity

Despite their varied origins, yoga, tai chi, and reiki share underlying principles that align with the concept of vitactions. These practices address the physical, mental, and emotional dimensions of health, fostering resilience against chronic diseases. They are rich in behavioral depth, combining mindfulness, intentionality, and physical activity—core elements that underpin their benefits. Though rooted in distinct cultural traditions, these practices have been widely adapted, offering accessible tools for addressing modern health challenges.

Integrating these types of activities into contemporary health paradigms requires a thoughtful approach. This includes tailoring practices to individual needs, from specific health conditions to lifestyle constraints, to ensure their relevance and effectiveness. Expanding research on their physiological and psychological mechanisms strengthens their credibility and adoption. Encouraging group participation amplifies their benefits, enhancing social connectivity and fostering a sense of community. Over the past couple of decades, beach areas across Europe have been heavily used in summer by groups practicing tai chi on the sand—showing that such activities can be embraced across diverse social and cultural contexts.

14.3 Conclusion

The wisdom of Asian traditions offers valuable insights into the prevention and management of chronic diseases through vitaction-like activities. Yoga, tai chi, and reiki exemplify how culturally rooted practices can address the deficiencies of modern lifestyles. By incorporating these practices into daily life, individuals can not only enhance their health but also reconnect with the fundamental principles of balance and harmony. This integration represents a promising avenue for addressing the growing burden of chronic diseases in an increasingly sedentary and stressed world.

The final two chapters explore the broader impact of vitactions—first by examining their societal implications, including how they influence health systems, cultural norms, and collective flourishing; then by uncovering their deeper potential as lifelong medicine for body, mind, and community.

Chapter 15. The Societal Implications of Vitactions

Vitactions are transformative not just for individuals but for entire societies. Although much of the discussion around vitactions has focused on personal health, their potential to reshape communities and address societal challenges is equally profound. Vitactions are inherently social, creating opportunities for connection, collaboration, and collective growth. By embracing them, it is feasible to address some of the most pressing issues of our time, from social isolation and chronic stress to declining mental health and disconnection in the modern age.

Societies today face unprecedented challenges. Rapid urbanization has fragmented communities, technological overuse has eroded face-to-face interaction, and many public spaces no longer prioritize the kind of activities that nurture connection and resilience. Vitactions offer a path forward. When embraced in public life, they spark ripple effects that uplift not only the individual but the community, strengthening connection, inspiring shared purpose, and helping life feel fuller, lighter, and more meaningful. This chapter explores the societal implications of vitactions, examining how they can be incorporated into public health strategies, urban planning, and global cultural practices to build a world enriched by connection and vitality.

15.1 Fostering Collective Well-Being

At their core, vitactions are about human connection—connection to ourselves, to others, and to the world around us. When these essential actions are practiced collectively, their impact is amplified, transforming communities into vibrant, supportive networks. A weekly walking group or a shared storytelling event is a simple initiative that can bring people together, fostering trust, empathy, and a sense of belonging. These shared activities become more than just moments of recreation; they are the building blocks of stronger, more resilient communities.

Consider a community that organizes regular cultural festivals where neighbors come together to share food, music, and stories. These gatherings provide more than entertainment; they become opportunities for individuals to connect, celebrate their heritage, and build relationships that sustain them through challenging times. Similarly, group activities like gardening in a communal space or participating in fitness or dance classes foster collaboration, promote physical health, and strengthen social bonds.

The benefits of collective vitactions are not just emotional or social; they extend to mental and physical resilience. Communities that prioritize shared activities are better equipped to cope with crises, whether they are natural disasters, economic disruptions, or public health emergencies. To maintain connection and support during the COVID-19 pandemic, some neighborhoods organized virtual storytelling sessions or outdoor exercise groups. These acts of collective resilience demonstrate the power of vitactions to sustain communities through adversity.

Across the globe, cultural traditions provide powerful examples of how shared vitactions can enrich societies. In Japan, the concept of "*moai*" involves forming lifelong social groups that provide emotional and financial support. This practice has been linked to lower rates of loneliness and longer life expectancy. In Nordic countries, the tradition of "*friluftsliv*", or outdoor life, encourages people to spend time in nature together, fostering physical health and emotional connection. These cases illustrate how vitactions can be embedded into cultural practices, creating opportunities for shared meaning, connection, and collective flourishing.

15.2 Vitactions in Public Health

Public health strategies have traditionally centered on medical interventions to address illness after it occurs. However, there is growing recognition of the role daily actions play in shaping health, leading to a shift toward more preventive, lifestyle-based approaches. Vitactions offer a unique and cost-effective means of promoting health and reducing the burden of chronic diseases. By addressing the behavioral roots of conditions like obesity, heart disease, anxiety and depression, vitactions create opportunities for lasting change.

Physical vitactions are powerful tools for improving cardiovascular health, reducing the risk of diabetes, and managing weight. These activities are accessible and adaptable, requiring little to no equipment, making them suitable for people across diverse socioeconomic backgrounds. Emotional vitactions like mindfulness practices, journaling, or engaging with music, help regulate mood, alleviate stress, and build emotional resilience. Social vitactions, including group meals, community events, or storytelling sessions, combat loneliness and foster a sense of belonging, addressing the growing mental health crisis.

The economic implications of integrating vitactions into public health strategies are substantial. Promoting preventive action through public campaigns and community programs is far more cost-effective than relying solely on medical treatments for conditions that could be avoided. A city that invests in safe walking paths and green spaces is likely to see long-term savings in healthcare costs related to sedentary lifestyles. Similarly, community programs that promote mindfulness and emotional support reduce the demand for mental health services and improve the quality of life.

Public health campaigns that emphasize vitactions can also inspire behavioral shifts on a larger scale. Educational initiatives that teach the benefits of physical activity, mindfulness, or storytelling can empower individuals to take control of their health. The most efficacious campaigns target specific populations, particularly older adults and underserved communities, to ensure that the benefits of interventions reach those who need them most.

15.3 Policy and Urban Planning

Policymakers hold the power to create environments that make vitactions accessible and appealing. Urban design has a profound influence on the opportunities people have for acquiring physical, social, and emotional vitactions. Cities that incorporate parks, walking paths, and shared spaces naturally encourage residents to take part in the kind of meaningful actions that enrich daily life.

A city that prioritizes walkable neighborhoods with interconnected green spaces, recreational facilities, and outdoor gathering areas encourages residents to walk, jog, or bike, promoting physical health and fostering opportunities for spontaneous social interactions. Parks become venues for mindfulness, exercise, and community events, enriching the lives of those who use them. Cultural hubs like libraries, art centers, and theaters provide spaces for storytelling, creative expression, and shared experiences, reinforcing social and emotional connection.

Workplaces and schools can also be designed to support vitactions. Flexible work schedules, wellness programs, and office layouts that encourage movement and collaboration can boost employee satisfaction, connection, and productivity. Schools that integrate play-based learning, group activities, and mindfulness exercises into their curricula lay the foundation for lifelong participation in vitactions.

Funding for public programs is another essential strategy. To ensure vitactions are accessible to all, governments must directly fund initiatives like fitness classes in parks, music therapy sessions in community centers, and intergenerational storytelling workshops. These investments not only improve individual health but also strengthen the social fabric of communities.

15.4 Global Perspectives on Vitactions

Cultural traditions around the world offer valuable insights into how vitactions can be seamlessly integrated into daily life. In many Indigenous cultures, storytelling is a central practice that conveys wisdom, strengthens community bonds, and fosters emotional connection. In fact, vitactions are rooted in the ways humans have always sought connection and meaning.

In India, yoga and meditation are deeply embedded in the culture, offering tools for physical and emotional balance that have stood the test of time. In Mediterranean societies, shared meals are a cherished tradition, providing opportunities for connection, celebration, and emotional nourishment. These practices highlight the universality of vitactions and their adaptability to diverse cultural contexts.

Global collaboration can amplify the impact of vitactions. By sharing best practices and innovative approaches, countries can learn from one another and create programs for promoting vitactions on a global scale. An international initiative that combines traditional wisdom could spark a new era of collective renewal and deeper connection.

15.5 Ethical and Practical Challenges

Despite their potential, ensuring equitable access to vitactions is a significant challenge. Resources such as parks, community centers, and welfare programs are often concentrated in wealthier areas, leaving marginalized populations with limited opportunities to participate in certain practices that should be considered essential. Policymakers must address these disparities to ensure that everyone has access to the benefits of vitactions.

Cultural sensitivity is another important consideration. Promoting vitactions must respect the traditions and values of different communities, avoiding a one-size-fits-all approach. Tailoring initiatives to local contexts ensures that they resonate with diverse populations.

Finally, balancing individual autonomy with public health goals is essential. Encouraging habits like walking, socializing, or mindfulness must be done in a way that empowers individuals rather than imposing prescriptive norms. Vitactions should be seen as opportunities for enrichment, not obligations.

15.6 Conclusion

The societal implications of vitactions are vast and transformative. By fostering connection, creativity, and resilience, these essential activities can address some of the most pressing challenges of our time. When integrated into public health strategies, urban planning, and cultural practices, vitactions have the power to create healthier, more connected societies.

As we embrace vitactions, we move toward a future where *joie de vivre* is not just a personal aspiration but a shared priority. By making these actions a central part of our lives, communities, and policies, we can build a world where everyone can thrive: a vitactin-rich society where collective flourishing becomes the norm.

Chapter 16. Vitactions as Medicine for Life

Vitactions provide the foundational support necessary for thriving, addressing gaps that modern lifestyles often leave unfilled. Vitamins combat physiological deficiencies; vitactins address behavioral deficiencies that can lead to stress, isolation, and mental health challenges. They are vital components of a balanced, fulfilling life.

As this book concludes, it is essential to reflect on the broader implications of vitactions. They hold the potential to transform individuals, communities, and societies. By integrating them into daily life, embracing them culturally, and embedding them in public policy, we can create a world where connection, creativity, and fulfillment are accessible to all. This chapter serves as both a synthesis of the ideas explored and a call to action, encouraging readers to make vitactions a cornerstone of their lives and to advocate for their widespread adoption.

16.1 The Power of Vitactions: From Personal to Global

The journey of vitactions begins with the individual. A single person's decision to incorporate physical movement, social interaction, or mindful reflection into their routine may seem small, but its impact can be profound. Take, for instance, someone who starts each day with a short walk, using the time to clear their mind and connect with nature. This simple act not only enhances their physical health but also uplifts their mood and sharpens their focus for the day ahead.

Yet the power of vitactions extends beyond the individual. When families, workplaces, and communities embrace these essential practices, their benefits multiply. Imagine a family that makes storytelling a nightly ritual. Parents and children not only share laughter and memories but also deepen their bonds and build a sense of belonging. Similarly, in a workplace that encourages group collaboration and movement breaks, employees experience reduced stress and increased productivity. These small changes, when scaled across communities, create a ripple effect that fosters stronger social networks and more resilient societies.

Vitactions also play a critical role in addressing some of the most pressing societal challenges. Loneliness and social isolation, often described as public health crises, can be mitigated through social vitactions like shared meals, group activities, or intergenerational storytelling. Meanwhile, learning new skills and engaging in creative pursuits are cognitive vitactions that help individuals maintain mental sharpness and emotional resilience in an increasingly complex world. By recognizing and prioritizing what truly matters, we cultivate not only healthier individuals but also more connected and compassionate communities.

16.2 Public Health and Policy Implications

The inclusion of vitactions in public health strategies has the potential to revolutionize healthcare systems and improve population-level outcomes. Unlike many traditional approaches that focus on treating illness after it arises, vitactions emphasize prevention and a more integrated approach to physical, mental, and social health. They address the root causes of many chronic conditions by encouraging daily practices that support physical, emotional, and social health.

Diseases ranging from diabetes and heart disease to depression are often linked to modifiable lifestyle factors, particularly physical inactivity, social disconnection, and chronic stress. As an accessible and cost-effective solution, engaging in vitactions directly counteracts these trends. Physical activities improve cardiovascular health, simultaneously releasing endorphins to reduce stress and boost mood. Meanwhile, practices like journaling and gratitude regulate mental health, and social vitactions strengthen the relationships that build a sense of community.

Policymakers have a vital role in creating environments that support vitactions. Urban design can prioritize walkability, green spaces, and recreational facilities, making it easier for people to take part in physical and social vitactions. Public funding for community centers, cultural events, and wellness programs ensures that these opportunities are accessible to people of all ages and socioeconomic backgrounds. Workplaces can adopt policies that encourage flexible schedules, allowing employees to integrate vitactions like movement breaks or mindfulness practices into their routines.

Healthcare providers can also incorporate vitactions into treatment plans, highlighting their value as complementary or alternative interventions. Combining mindfulness practices with pharmacological treatment for anxiety or recommending storytelling sessions for older adults with memory difficulties, can enrich conventional care and support a more holistic approach that addresses the whole person.

16.3 Cultural and Global Perspectives

Vitactions are not a new concept; they are deeply embedded in the traditions and practices of cultures around the world. Indigenous communities have long understood the importance of connection to nature, storytelling, and communal rituals for maintaining balance. In many Asian cultures, mindfulness and meditation have been practiced for centuries as essential tools for managing stress and fostering emotional stability. European traditions often celebrate art, music, and shared meals as central to a fulfilling life.

These cultural practices highlight the universality of vitactions, even as their specific expressions vary. They remind us that walking, singing, and storytelling are not merely recreational, they are fundamental expressions of what it means to be human. By drawing on these traditions, we can enrich our understanding of vitactions and find new ways to integrate them into contemporary life.

Global collaboration also holds great promise for expanding the reach and impact of vitactions. By integrating traditional practices with modern scientific research, we could build a shared, evidence-informed blueprint for promoting vitactions worldwide. Such an effort could cultivate cross-cultural understanding and spark fresh approaches to deepening the human experience and nurturing shared growth.

16.4 Call to Action

Vitactions are not just a concept—they are a movement, a call to action for individuals, communities, and societies to prioritize meaningful actions that enrich daily life and strengthen our shared human experience. As a reader, you have the power to start this transformation in your own life. Begin with small, intentional changes: take a walk every day, share a story with a loved one, or listen to a piece of music that inspires you. These simple acts, when practiced consistently, have the potential to reshape not only your own health but also the world around you.

Communities and policymakers must also rise to the challenge. By investing in public spaces, cultural programs, and educational initiatives, they can create environments where vitactions thrive. Workplaces, schools, and healthcare systems can incorporate vitactions into their daily operations, making them an integral part of modern life.

Beyond these practical steps lies a deeper realization: that what we do each day is as vital to our lives as the food we eat or the remedies we take. This realization demands a cultural change, a reimagining of what it means to live well. Vitactions remind us that health is not just the absence of illness but the presence of connection, creativity, and joy. They challenge us to think beyond individual success and embrace the collective flourishing of humanity.

16.5 Closing Reflections

Vitactions are more than a guide for individual action—they represent a philosophy, a way of seeing life as a tapestry of meaningful acts that connect us to ourselves, to others, and to the world. Like vitamins, they address deficiencies that may not be outwardly visible but are deeply felt. In an age that often prioritizes speed over depth and productivity over presence, they offer a path toward resilience, balance, and a more fulfilling life.

As this book comes to a close, let the concept of vitactions inspire you to reimagine your own life. Whether through a morning walk, a heartfelt conversation, or a creative project, these simple yet meaningful actions have the power to transform not only your days, but also the legacy you leave behind. By embracing vitactions, we move toward a world where joy, connection, and a sense of meaning are not rare exceptions, but everyday realities.

EPILOGUE

What *You Saw*... and What *You Saw* ***Without Knowing*** **It**

If you've made it this far, you already know: the images accompanying this book weren't just illustrations. They were little tests of attention. Seeds of **vitactions** scattered throughout the chapters.

Did you notice anything amiss? You may have noticed subtle errors: misspelled or incomplete words, or images that didn't quite make sense. Maybe you didn't pinpoint the errors, but a feeling lingered that something in this reality was just a little bit broken.

That was exactly what you were meant to experience.

The images weren't designed to deceive you, but to invite you to look. To stop automatic reading. To activate the conscious perception we so often neglect. In other words: they were practical training for the same idea that runs throughout the book.

If you noticed the errors, your mind performed a **vitaction**.

If you didn't notice them at first, but something compelled you to return, you did that too.

If you wondered why they happened, even better.

Vitactions aren't just learned; they're practiced.

And if this book has made your gaze a little more attentive, a little more curious, and a little more alive, then it has fulfilled its purpose.

Thank you for reading, but above all, thank you for *looking*.

Declaration: The images in this book were created using *Sora* during a period when it was available as a tool within ChatGPT (January 2025).

Appendices

Appendix A: Glossary of Terms

This glossary defines key terms introduced throughout the book, ensuring clarity and consistency for readers.

- **Vitactions/Vitactins** are defined as behaviors essential for maintaining emotional, cognitive, and social well-being, similar to how vitamins are necessary for physical health.
- A **vitaction/vitactin** denotes an individual, specific action within that behavioral framework.

Vital Actions: Vitactions

Behavioral Deficiency Syndrome: A hypothetical state where the lack of essential vitactions leads to physical, emotional, and cognitive dysfunctions, akin to the effects of vitamin deficiencies in the body.

Behavioral Rickets: a term proposed by the author, encapsulates the consequences of a sedentary lifestyle.

Behavioral Scurvy: a term proposed by the author, describes the effects of insufficient social engagement.

Behavioral Stagnation: term proposed by the author, describes the effects of insufficient cognitive stimulation

Cognitive Vitactions: Actions that stimulate the brain, enhance neuroplasticity, and promote learning and memory, such as puzzles, language learning, or creative writing.

Emotional Vitactions: Behaviors that support emotional balance and stress management, including mindfulness, gratitude practices, or listening to music.

Physical Vitactions: Activities that promote movement, physical health, and vitality, such as walking, yoga, or dancing.

Social Vitactions: Actions that foster connection and community, such as storytelling, group activities, or shared meals.

Vitaction-Enriched Environment: Spaces or settings designed to naturally encourage and support the practice of vitactions, such as parks, community centers, or walkable neighborhoods.

Annotated Bibliography of Key References

This annotated bibliography highlights foundational works and recent studies that inform the concept of **vitactions**.

- Watve, M., and Keskar Sardeshmukh, A. (2024). "Vitaction deficiency: a possible root cause for multiple lifestyle disorders including Alzheimer's disease."[1]

This article introduces the term "**vitaction**" and draws parallels between behavioral and vitamin deficiencies. The authors propose that certain essential behaviors, called vitactions, are crucial for maintaining physiological and neurological health. They suggest that the absence of these behaviors may contribute to lifestyle-related disorders, such as Alzheimer's disease, by linking vitactions to brain health through mechanisms such as angiogenesis and mitochondrial function.

- Franco, R. (2024). "Vitactions: vitamins for the brain."[2]

This opinion piece expands on the concept of vitactions, highlighting their role in the prevention of neuropsychiatric and degenerative diseases. It proposes a classification framework for vitactions, emphasizing their importance throughout life. To further exploring its potential for research and social application, the article also introduces the term "**vitactin**" as an individual type of vitaction. A vitactin would be synonym of a vitaction.

The references described below have been integrated to support the fundamental concepts and practical implications discussed throughout the book.

- Lam et al., (2021). "Neurobiology of loneliness: a systematic review " [3]

This review compiles studies on the neurological impact of social interaction and its role in mental health, providing evidence on the importance of social vitactins.

- Zhamaliyeva et al., (2025). " Interventions to Reduce Mental Health Stigma Among Health Care Professionals in Primary Health Care: A Systematic Review and Meta-Analysis" [4]

- Wakabayashi et al., (2023). "Psychological aspects of rehabilitation nutrition: A position paper by the Japanese Association of Rehabilitation Nutrition" [5]

Reviews on the role of mindfulness in managing stress, anxiety, and chronic conditions, demonstrating the power of emotionally driven actions.

- Aboelata M.J., and Navarro, A.M. (2010) "Emerging issues in improving food and physical activity environments: strategies for addressing land use, transportation, and safety in 3 California-wide initiatives" [6]

- Mmari et al., (2025) “Cultivating Health: The Role of Urban Greening in Supporting Baltimore's Youth” [7]

These articles describe how environmental planning can influence physical activity, social connectivity, and overall well-being.

- Zhang et al., (2021). " Musical Experience Offsets Age-Related Decline in Understanding Speech-in-Noise: Type of Training Does Not Matter, Working Memory Is the Key." [8]

- Ting et al., (2024) “The Sound of Memory: Investigating Music Therapy’s Cognitive Benefits in Patients with Dementia—A Network Meta-Analysis of Randomized Controlled Trials” [9]

These studies highlight the cognitive and emotional benefits of music-based activities in elderly populations.

References

[1] M. Watve and A. Keskar Sardeshmukh, "'Vitaction' deficiency: A possible root cause for multiple lifestyle disorders including Alzheimer's Disease," *Explor. Neuroprotective Ther.*, vol. 4, pp. 108–118, 2024, doi: 10.37349/ENT.2024.00074.

[2] R. Franco, "Vitactions: vitamins for the brain," *Explor. Neuroprotective Ther.*, vol. 4, no. 4, pp. 300–307, Jul. 2024, doi: 10.37349/ENT.2024.00084.

[3] J. A. Lam *et al.*, "Neurobiology of loneliness: a systematic review," *Neuropsychopharmacology*, vol. 46, no. 11, pp. 1873–1887, Oct. 2021, doi: 10.1038/S41386-021-01058-7.

[4] L. Zhamaliyeva *et al.*, "Interventions to Reduce Mental Health Stigma Among Health Care Professionals in Primary Health Care: A Systematic Review and Meta-Analysis," *Int. J. Environ. Res. Public Health*, vol. 22, no. 9, Sep. 2025, doi: 10.3390/IJERPH22091441.

[5] H. Wakabayashi *et al.*, "Psychological aspects of rehabilitation nutrition: A position paper by the Japanese Association of Rehabilitation Nutrition," *J. Gen. Fam. Med.*, vol. 25, no. 1, pp. 1–9, Jan. 2023, doi: 10.1002/JGF2.668.

[6] M. J. Aboelata and A. M. Navarro, "Emerging issues in improving food and physical activity environments: strategies for addressing land use, transportation, and safety in 3 California-wide initiatives," *Am. J. Public Health*, vol. 100, no. 11, pp. 2146–2148, Nov. 2010, doi: 10.2105/AJPH.2010.193466.

[7] K. Mmari *et al.*, "Cultivating Health: The Role of Urban Greening in Supporting Baltimore's Youth," *J. Urban Health*, Nov. 2025, doi: 10.1007/S11524-025-01022-6.

[8] L. Zhang, X. Fu, D. Luo, L. Xing, and Y. Du, "Musical Experience Offsets Age-Related Decline in Understanding Speech-in-Noise: Type of Training Does Not Matter, Working Memory Is the Key," *Ear Hear.*, vol. 42, no. 2, pp. 258–270, Mar. 2021, doi: 10.1097/AUD.0000000000000921.

[9] B. Ting *et al.*, "The Sound of Memory: Investigating Music Therapy's Cognitive Benefits in Patients with Dementia—A Network Meta-Analysis of Randomized Controlled Trials," *J. Pers. Med.*, vol. 14, no. 5, p. 497, May 2024, doi: 10.3390/JPM14050497/S1.

www.ingramcontent.com/pod-product-compliance
Lightning Source LLC
LaVergne TN
LVHW022107200926
848481LV00003B/1090
* 9 7 9 8 8 9 9 6 6 3 6 0 4 *